1-17-97

AN INTRODUCTION TO
STATISTICAL
THERMODYNAMICS

AN INTRODUCTION TO
STATISTICAL
THERMODYNAMICS

Robert P H Gasser & W Graham Richards

Oxford University

World Scientific
Singapore • New Jersey • London • Hong Kong

Published by

World Scientific Publishing Co. Pte. Ltd.

P O Box 128, Farrer Road, Singapore 9128

USA office: Suite 1B, 1060 Main Street, River Edge, NJ 07661

UK office: 57 Shelton Street, Covent Garden, London WC2H 9HE

Library of Congress Cataloging-in-Publication Data

Gasser, R. P. H. (Robert Paul Holland)
    An introduction to statistical thermodynamics / Robert P.H. Gasser
& W. Graham Richards.
      p.   cm.
     ISBN 9810222432
     ISBN 9810223722 (pbk)
     1. Statistical thermodynamics.   I. Richards, W. G. (William
Graham) II. Title.
    QC311.5.G63   1995
    536'.7--dc20   95-20097
                                   CIP

British Library Cataloguing-in-Publication Data
A catalogue record for this book is available from the British Library.

This book is printed on acid-free paper.

Printed in Singapore by Uto-Print

PREFACE

As College Tutors we find that a major difficulty in teaching statistical thermodynamics to chemistry students is that (somewhat surprisingly) when they first encounter this subject they often find the underlying physical principles more difficult to grasp than the mathematical formalism. Thus the question 'What is a partition function?' is usually met with blank incomprehension, even though the equation defining this function is well-known and its derivation can be reproduced on request. A number of existing textbooks provide excellent coverage of the more formal aspects of the subject and examine the philosophical foundations on which it rests. In this book, however, one of our principle aims is to make clear to the beginner the physical basis of the relation between the observable behaviour of systems of atoms or molecules and the properties of the individual particles. Some acquaintance with elementary calculus and quantum theory is assumed, as is a familiarity with the basic ideas of classical thermodynamics.

In pursuing our aim we have occassionally sacrificed rigour for the sake of clarity and we have thought it especially important to introduce at the earliest possible moment the world of experiment. This we regard as the best way to an understanding of the subject. We have chosen therefore to illustrate the principles by including reference to some modern experimental subjects such as lasers, spectroscopy, and superconductivity. It is our hope that, with this approach, even the student new to the subject will be enabled to see the immediate relevance of statistical thermodynamics to a number of subjects, which he will quickly recognize as being of general scientific interest and importance. We also hope that this attempt to provide a more widely scientifically cultured approach than is usual in elementary books will offer some insight into the powerful and exciting pervasiveness of the ideas of entropy and energy levels in atomic and molecular systems. The book does not attempt to provide a comprehensive coverage of the subject and, in particular, we have thought it wise to exclude a full account of the statistical mechanics of solutions and of polymers. Solutions are treated briefly but any worthwhile discussion of polymers would have required a considerable expansion to do justice to their importance. In our view this would have been undesirable. Our preference for a compact book is also reflected in the choice and number of problems. Those we have put in are intended more to illustrate and extend the discussion in the text, and thereby to increase the student's 'feeling' for the subject, than to test his ability to do numerical calculations.

The first chapters provide a basic outline of the subject and illustrate the influence of quantum effects. The ideas are then applied to a number of subjects, in which we hope that the reader will share our interest. Finally, we have allowed ourselves to range outside the more conventional bounds of the subject into areas of human endeavour where the applicability of statistical ideas is not yet fully established—and perhaps never will be. This may be thought a trifle over-ambitious in an elementary book, but it is part of our second important aim in writing this book, which has been to stimulate interest; even where, in so doing, the discussion may offend the purist.

We wish to express our thanks to our colleagues in the Physical Chemistry Laboratory with whom we have had many discussions about the contents of this book. These discussions have always been stimulating, often controversial, and usually enlightening.

Since first published in 1974 as 'Entropy and Energy Levels', the book has been very popular with students and we hope that this revised and updated version will serve the same needs.

Oxford 1995

ACKNOWLEDGEMENTS

We acknowledge gratefully permission to reproduce material from the following sources:

FIG. 5.3(a): H. Wise, *J. Phys. Chem.* **58**, 389 (1954). FIG. 5.3(b): R. N. Doescher, *J. Chem. Physics*, **20**, 331 (1952). FIG. 5.4: R. F. Barrow and J. G. Stamper, *Trans. Faraday Soc.*, **54**, 1592 (1958). Appendices 1, 2, and 3: J. E. Mayer and M. G. Mayer, *Statistical Mechanics*, Wiley, New York, 1940.

ACKNOWLEDGMENTS

CONTENTS

Chapter 1

THE LAWS OF THERMODYNAMICS

THERMODYNAMICS is a subject that seeks to interpret the properties of chemical systems. It is particularly concerned with systems at equilibrium and no attention is paid to the rate at which equilibrium is achieved: this latter subject is the province of reaction kinetics. So far it has been found that three laws of thermodynamics are necessary and sufficient. An understanding of the first two laws does not need an atomic theory of the nature of matter though the laws can be appreciated much better with such a theory. The third law can be comprehended properly only from a microscopic molecular viewpoint. Of these laws, it is with the third that we shall be principally concerned. But, because of the close inter-relation of all three, we start with a brief consideration of each.

1.1. The First Law of Thermodynamics

The First Law is the well known energy-equivalence law. 'Energy cannot be created nor destroyed, only converted from one form to another'. The most useful algebraic formulation of this law for the chemist is

$$q = \Delta U - w,$$

where q is a quantity of heat supplied to the system, ΔU is the change in internal energy and w is any work done on the system by external forces.

Classical thermodynamicists felt no need of statistical pictures to explain this law nor to interpret the thermodynamic functions, q, U, and w.

Nonetheless statistical concepts do aid our understanding. Particularly we think of the internal energy U as the sum of the energies of the components. To take the familiar case of a monatomic ideal gas: the internal energy is simply the sum of the kinetic energies of the component atoms. When an infinitesimal quanitity of heat δq is supplied to an ideal gas there may be an increase in internal energy δU and work may be done by the gas if it is allowed to expand by a volume δV. Under normal laboratory conditions, where the external pressure is kept constant, the work done by the gas $(= -\delta w)$ is $P\delta V$. Thus

$$\delta q = \delta U + P\delta V$$

or, for a finite change from an initial state A to a final state B,

$$q_B - q_A = U_B - U_A + P(V_B - V_A) = (U_B + PV_B) - (U_A + PV_A).$$

It is convenient to have a new thermodynamic function for $(U + PV)$, which is called the *enthalpy, H*. This function is appropriate when we are considering changes at constant pressure.

1.2. The Second Law of Thermodynamics

A traditional way of expressing the second law is in terms of heat engines. This seems a long way from the behaviour of molecules although mathematically the connection can be made quite readily. From a consideration of the properties of heat engines operating reversibly, it is possible to define a new thermodynamic function, the entropy, S, which is governed by the equation

$$\Delta S = \frac{q_{\text{reversible}}}{T}.$$

A readily measurable entropy change occurs when a pure substance changes phase. In this case $q_{\text{reversible}}$ and T are respectively the molar latent heat, L (e.g. of melting or boiling) and the temperature of the phase change. It has been observed that the entropy change on boiling (ΔS_b) many non-associated organic liquids is approximately constant:

$$\Delta S_b = \frac{L}{T_b} \approx 86 \text{ J K}^{-1}\text{mol}^{-1}.$$

This value of ΔS_b is often called Trouton's constant.

Hereafter we shall omit the subscript 'reversible'; all changes are assumed to be carried out reversibly. For molecular problems the clearest statement of the second law is *When left to themselves systems tend to alter in such a way that the entropy of the system increases.* The understanding of this statement is helped by an understanding of the concept of entropy. Statistically, entropy is thought of as a measure of the chaos or disorder of the system or even, less exactly, as the 'ignorance' about a situation. This latter rather qualitative statement will be returned to at the very end of this book, whilst in Chapter 3 a firm analytical definintion of 'disorder' will be given. Even in an undefined way the idea that entropy, as measured by disorder, tends to increase spontaneously is in accord with everyday experience and enables us to rationalize directions of change.

The position of a chemical equilibrium at constant temperature is governed by two factors. While the energy tends to reach a minimum the entropy tends to reach a maximum. The balance between these two, often opposing, factors is described by another thermodynamic function, free energy. Depending on whether the system is at constant volume or at constant pressure the free energies are defined as follows:

$$A = U - TS \qquad \text{Helmholtz free energy (constant volume)}.$$

$$G = H - TS \qquad \text{Gibbs free energy (constant pressure)}.$$

The second law, however, gives us no information about the absolute values of S, A, and G but only about *changes* in these quantities.

1.3. The Third Law of Thermodynamics

The third law of thermodynamics is concerned with the behaviour of the absolute value of entropy as the absolute zero of temperature is approached. Although the first and second laws do not rely on statistical interpretations of the thermodynamic functions, a proper understanding of the third law is possible only if we provide a microscope molecular interpretation. Various formulations of this law have been used and we shall use the phraseology of Simon:

'The contribution to the entropy of a system by each aspect which is in internal thermodynamic equilibrium tends to zero at the absolute zero'.

Familiar examples of 'aspects' include the rotational and vibrational motions of molecules, the process of mixing, and the spatial disposition of the magnetic moments of unpaired electrons in paramagnetic compounds. The consequence of the third law is that the total entropy of a substance tends to zero at the lowest temperatures and thus that all these properties should achieve an ordered state as the absolute zero is approached. However, the third law gives no guidance about the mechanism by which the ordering is to be achieved. It is worthwhile comparing our definition of the third law with another familiar formulation:

'The entropy of a perfect crystalline solid tends to zero as the temperature tends to the absolute zero'.

A 'perfect crystalline solid' is thus the analogue of 'an aspect in internal thermodynamic equilibrium'.

In making an experimental confirmation of the third law one often measures the heat capacity and calculates the entropy of a substance from as low a temperature as possible up to room temperature, and then compares the result with the entropy calculated from the formulae of statistical thermodynamics. In many cases there is excellent agreement between theory and experiment and there are good grounds for believing that the third law is as valid as the first and second laws. Also intersting are those few cases in which there are discrepancies between theory and experiment. These discrepancies are always in the direction of the system having more entropy than measurements based on the third law appear to indicate. Such discrepancies suggest that even at the lowest temperatures the sample retains some disorder; i.e., for at least one aspect of the system the entropy has not vanished. The reason for a non-vanishing entropy is that the aspect with which the entropy is associated has not maintained its true internal thermodynamic equilibrium. This failure to maintain equilibrium need not surprise us, since it is commonplace in everyday life; mixtures of reactive chemicals whose equilibrium composition overwhelmingly favours products which are produced in a strongly exothermic reaction frequently fail to react (fortunately for the motorist) in the absence of an external agent. Similarly, when an aspect of a system has reached a low enough temperature for its entropy to tend to zero there may be no kinetic pathway of sufficiently low energy available. For example, the two isotopes of chlorine that go to make up sodium chloride crystal have an entropy of mixing that should disappear at low temperatures. However, this ordering process would require the movement of chloride ions through the lattice, which is impossibly slow at low temperatures, and so the entropy remains. Anomalies in absolute entropies are often associated with the difficulty a constituent of a solid may have in undertaking molecular motion at low temperatures. However it is necessary to consider each particular example of an anomaly separately in order to identify in detail the origin of the residual entropy. We shall discuss this subject in more detail in Chapter 10.

1.4. Some Useful Results from Classical Thermodynamics

Having summarized the laws of thermodynamics we now give an even briefer résumé of the results of classical thermodynamics. These are quite simply obtained starting with the first two laws and the definitions of U, H, S, A, and G.

Two *heat capacities* are commonly defined. These are the heat capacity at constant volume (C_V) and the heat capacity at constant pressure (C_P):

$$C_V = \left(\frac{\partial U}{\partial T}\right)_V \text{ and } C_P = \left(\frac{\partial H}{\partial T}\right)_P.$$

For closed systems the following set of equations expresses the changes in the thermodynamic functions when the only work that can be done is *PV* work:

$$dU = TdS - PdV$$

$$dH = TdS + VdP$$

$$dA = -PdV - SdT$$

$$dG = VdP - SdT.$$

If we now suppose that the system is further partially restricted by putting one of the differentials at a time equal to zero we can obtain the following set of equations:

$$\left(\frac{\partial U}{\partial S}\right)_V = T \qquad \left(\frac{\partial U}{\partial V}\right)_S = -P$$

$$\left(\frac{\partial H}{\partial S}\right)_P = T \qquad \left(\frac{\partial H}{\partial P}\right)_S = V$$

$$\left(\frac{\partial A}{\partial T}\right)_V = -S \qquad \left(\frac{\partial A}{\partial V}\right)_T = -P$$

$$\left(\frac{\partial G}{\partial T}\right)_P = -S \qquad \left(\frac{\partial G}{\partial P}\right)_T = V \, .$$

From these relations we can derive a number of useful results.

The first relation gives us

$$\left(\frac{\partial S}{\partial U}\right)_V = \frac{1}{T} \, .$$

This we will use in Chapter 3. Putting $(\partial G/\partial T)_P = -S$ into the definition of G gives us the *Gibbs-Helmholtz relation*

$$H = G - T\left(\frac{\partial G}{\partial T}\right)_P \, .$$

Further, the relation $(\partial G/\partial P)_T = V$ can lead us to the *Gibbs isotherm*. If we substitute the ideal-gas equation for 1 mol $(PV = RT)$ in the relation $(\mathrm{d}G)_T = V\mathrm{d}P$ we get:

$$(\mathrm{d}G)_T = \frac{RT \, \mathrm{d}P}{P} = RT \, \mathrm{d} \ln P$$

or

$$\Delta G = RT \ln P_2/P_1 \, .$$

We define the standard free energy $G^\ominus$ as the value of G for 1 mol of gas at 1 atm; then

$$G - G^\ominus = RT \ln\frac{P}{1} = RT \ln P \, .$$

For a gaseous reaction:

$$a\mathrm{A} + b\mathrm{B} \rightleftharpoons c\mathrm{C} + d\mathrm{D}$$

where the pressures are P_A, P_B, P_C, and P_D we can write for each component:

$$G_i = G_i^\ominus + RT \ln P_i$$

so that the overall free-energy change is

$$\Delta G = \Delta G^\ominus + RT \ln\left(\frac{P_C^c P_D^d}{P_A^a P_B^b}\right)$$

where

$$\Delta G^\ominus = cG_C^\ominus + dG_D^\ominus - aG_A^\ominus - bG_B^\ominus \, .$$

If the system is at equilibrium $\Delta G = 0$ and

$$\Delta G^{\ominus} = -RT \ln K_P \, (the \; Gibbs \; isotherm) \, .$$

This on substitution into the Gibbs-Helmholtz relation then yields the useful *Van't Hoff isochore*:

$$\frac{\mathrm{d} \ln K_P}{\mathrm{d}(1/T)} = -\frac{\Delta H^{\ominus}}{R} \, .$$

These results will be familiar to anyone who has studied elementary thermodynamics but are collected here for the convenience of the reader.

Having thus quickly reviewed some classical thermodynamics we are in a position to set out to understand statistical thermodynamics. The line of reasoning we shall follow is that all macroscopic properties such as energies, entropies, or free energies must ultimately depend on the properties of the molecules which make up the system. While we can, in principle, calculate the energy levels available to the particles that constitute the system, this information is not by itself sufficient to allow us to determine the thermodynamic properties. These, we shall see, are strongly dependent on the way in which the molecules are distributed among the energy levels. One of our aims is to show how it is possible to relate the characteristic properties of atoms and molecules, which form part of an ensemble, to the observations we can make, which are of necessity on the system as a whole. In particular we shall be much concerned with the energy and the entropy of pure substances, properties which depend on how the molecules are distributed among their available energy levels. This distribution depends in turn both on temperature and on the intimate nature of the molecules.

Chapter 2

DISTRIBUTION LAWS

The bulk properties of an assembly of particles necessarily depend on the properties of the individual, microscopic components. The problem lies in determining the form of the dependence. In order to do this, which is one of our central purposes, we shall use as a model a system in which the individual particles interact with each other only enough to maintain proper thermodynamic equilibrium; then at constant volume the total energy of the system is independent of the positions of the particles. This model represents an idealized situation, as can be seen if we remember that even the inert gases have attractive forces between atoms, forces which eventually lead to liquefaction and solidification. The particles of this model are sometimes described as 'quasi-independent'.

In thermodynamics we are particularly interested in energy and entropy. Using our model, the internal energy of a gas can be taken as the sum of the energies of the individual molecules. Classical statistical thermodynamics placed no limits on the values which the internal energies could have. A diatomic molecule, for example, could move, rotate, or vibrate without restriction. With the advent of quantum mechanics, however, it has become clear that there are severe restrictions on the rotatory and vibratory motions of molecules, with only a limited number of energy states being accessible. Translational motion, on the other hand, although quantized, has such small separations between energy levels that quantization produces no measurable difference from classical behaviour. Our aim now is to investigate the way in which non-reacting particles distribute themselves among their available energy levels. In particular, with the system maintained at constant temperature (and therefore constant energy) *we should like to know the number of molecules having each value of the permitted energies.* Armed with this knowledge of the distribution of the particles among their energy levels we shall then be able to calculate the bulk thermodynamic properties of the system. This determination of the macroscopic properties of an assembly of particles from the properties of the individual particles is a fundamental part of statistical thermodynamics.

In normal systems at temperatures above the absolute zero of temperature the tendency of particles to escape from the ground state, which they occupy at the absolute zero, causes them to become distributed in numerous different ways amongst the energy levels. In statistical thermodynamics we set out to find the most probable of these distributions and then to relate this to the bulk thermodynamic functions. It is remarkable and fortunate that we need to deal only with the most probable distribution and that deviations from it, for the large numbers of molecules with which a chemist deals, can be ignored. Before discussing the distributions, however, we ought to say a few words about energy levels.

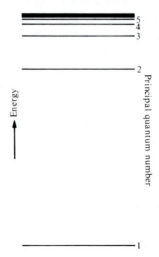

Fig. 2.1. Energy levels of the hydrogen atom.

Energy Levels

The earliest results of quantum theory showed that confined particles could not have a continuous range of energy but rather that only certain specific energies are allowed. The hydrogen atom was shown by Bohr to have the well known set of energy levels (Fig. 2.1).

The energy of any atomic level is given by the formula

$$E_n = -R_\infty/n^2 \,,$$

where n is an integer and R_∞ the Rydberg constant. As well as electronic energy, molecules can have vibrational energy, which is also 'quantized'; for a diatomic molecule we can draw the potential curve (Fig. 2.2) where the horizontal lines indicate the allowed quantum levels of vibrational energy.

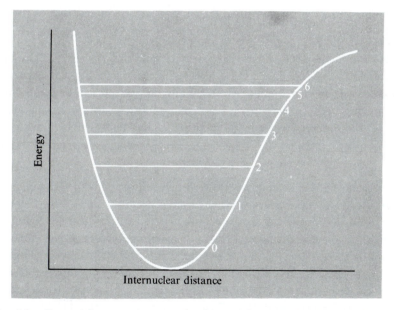

Fig. 2.2. Potential-energy curve and vibrational energy levels of a diatomic molecule.

The rotational motion of molecules is also governed by quantum conditions and for a diatomic molecule the energy of any rotational level J is given by

$$E_J = \frac{h^2}{8\pi^2 I} J(J+1)$$

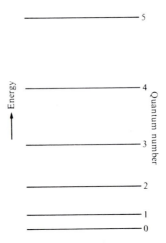

Fig. 2.3. Energy levels of a rigid rotor.

where I is the moment of inertia of the molecule and J is any positive integer or zero. The spacing of these levels is illustrated in Fig. 2.3. Although translational energy is quantized, for many purposes it can be treated as a continuum, because the separations of the energy levels are very small.

All these energy levels can, in principle, be obtained by solving the Schrödinger equation

$$H\Psi = E\Psi$$

where H is the Hamiltonian operator (which corresponds to the sum of the kinetic and potential energies of the system) and Ψ is the wave-function. The resulting values of E are the allowed energies. If the system contains a number of non-interacting particles the wave-function for the overall system would be the product of the wave-functions for all the particles. Thus for two particles, 1 and 2, with wave functions Ψ_a and Ψ_b a first estimate of the total wave-function might be $\Psi_a(1)\Psi_b(2)$. This, however, turns out not to be a satisfactory description (if the particles are identical) for it has been discovered that Nature puts a further constraint on the wave-functions of many-particle systems. When any two identical particles are interchanged the wave-function must either remain unchanged (the wave-function is then said to be *symmetric*) or, at most, change its sign (be *antisymmetric*). The square of the wave-function must on no account become different. If it were to do so, this would imply that a physically different situation has been produced by the interchange of identical particles.

To take account of these symmetry requirements we could write the total wave-function for our two particles as $\Psi_a(1)\Psi_b(2) + \Psi_a(2)\Psi_b(1)$ (symmetric) or $\Psi_a(1)\Psi_b(2) - \Psi_a(2)\Psi_b(1)$ (antisymmetric). We shall say more about this later but we must notice here that these restrictions on the wave-functions may affect the distributions of particles in their energy levels. There are thus two quantum-mechanical situations to be discussed; particles with symmetric wave-functions (often called *bosons*) and particles with anti-symmetric wave-functions (often called *fermions*). Fortunately, under most (but not all) circumstances of interest to chemists the two distributions become identical and, moreover, the distribution of particles is the same as the distribution calculated without reference to the symmetry of the wave functions. In fact, much the most important and widely used distribution law is the classical law of Boltzmann, which can be derived without reference to symmetry considerations. We shall, therefore, derive this law first and then show, briefly, how symmetry effects modify the result.

Boltzmann Distrbution Law

This is the classical case in which we make no symmetry restrictions. The total wave-function is then simply the product of the wave-functions for the individual constituent particles; i.e.

$$\Psi_{total} = \Psi_a(1)\Psi_b(2)\ldots\Psi_m(n).$$

The numbers $1, 2\ldots n$ label the particles and the letters $a, b\ldots m$ label the wave-functions. We start by considering the situation in which the particles are localized and so can be treated as distinguishable by virtue of their spatial positions. If we exchange any two, e.g. 1 and 2, we have a new 'state' of the system:

$$\Psi'_{total} = \Psi_a(2)\Psi_b(1)\ldots\Psi_m(n).$$

We want to answer the question 'What is the most probable distribution of particles among the available energy levels for a particular, fixed value of the total energy?' First, however, we need to answer a simpler question which is 'How many ways can we distribute N particles in a certain fixed number of energy levels?' This is the same question as 'How many total wave-functions are there that correspond to a distribution of N particles, such that there are n_1 particles of energy ε_1, n_2 particles of energy ε_2, n_i particles of energy ε_i etc.?'

Let us first consider the simple case where there are two energy levels ε_1 and ε_2. If $N = 1$, the possibilities are $n_1 = 0$ and $n_2 = 1$, or $n_1 = 1$ and $n_2 = 0$; i.e. W, the number of ways of distributing the particles, is 2. This can be extended for other values of N:

N	n_1	n_2	W	
1	1	0	1 ⎫	
	0	1	1 ⎭	2 ways
2	0	2	1 ⎫	
	1	1	2 ⎬	4 ways
	2	0	1 ⎭	
3	0	3	1 ⎫	
	1	2	3 ⎬	
	2	1	3	8 ways
	3	0	1 ⎭	
4	0	4	1 ⎫	
	1	3	4	
	2	2	6 ⎬	16 ways
	3	1	4	
	4	0	1 ⎭	

And so, for a two-level system

$$W_{2-\text{level}} = \frac{N!}{n_1! n_2!} . \qquad (\text{remember } 0! = 1)$$

In general, with many levels,

$$W = \frac{N!}{n_1! n_2! n_3! \ldots} .$$

In order to find the most probable way in which the particles will be distributed we shall need to assume that the intrinsic probabilities of all states of the system are equal. This postulate is of fundamental importance and implies that Nature has no predilection for any particular type of state, an inherently reasonable but nonetheless unprovable proposition. The justification for this assumption lies essentially in the agreement of our final equations with experimental observations. It follows from the postulate that the probability, P, of any particular distribution of particles is directly proportional to the number of ways in which that distribution can be realized:

$$\text{i.e. } P \propto W.$$

It is our concern, therefore, to find the values of the various n_i's which make P a maximum, subject to two conditions: (i) that the total number of particles remains constant, and (ii) that the total energy of the system E remains constant. Clearly W can increase indefinitely if these two conditions are not imposed. When P is a maximum any small change in the distribution of the particles leaves P unchanged, i.e. $\delta P = 0$, and conditions (i) and (ii) require that $\delta N = 0$ and $\delta E = 0$. Now $\ln P$ is also a maximum when P is a maximum and it is more convenient mathematically to work with $\ln P$ than with P. Thus we now seek the values of the n_i's which will make $\ln P$ a maximum.

Any small change in $\ln P$ can be expressed by

$$\delta \ln P = \frac{d(\ln P)}{dn_1}\delta n_1 + \frac{d(\ln P)}{dn_2}\delta n_2 + \dots$$

but for a maximum

$$\delta \ln P = 0 \tag{2.2}$$

and we also have the constraints

$$\delta N = \sum_i \delta n_i = 0 \tag{2.3}$$

$$\delta E = \sum_i \varepsilon_i \delta n_i = 0 \tag{2.4}$$

since

$$N = \sum_i n_i \quad \text{and} \quad E = \sum_i n_i \varepsilon_i.$$

We require all three conditions, Eqs. (2.2), (2.3), and (2.4) to be fulfilled simultaneously and we therefore need a mathematical technique which incorporates this requirement in our expression for the values of n_i which make ln P a maximum. The required technique in Lagrange's method of undetermined multipliers, in which we multiply the second and third of these equations by constants α and β respectively, which may be evaluated later and combine the three relationships into a single equation, i.e.

$$\sum_i \left(\frac{d \ln P}{dn_i} + \alpha + \beta \varepsilon_i \right) \delta n_i = 0 \,.$$

Now if the arbitrary and independent δn_is are not zero, then their coefficients must be zero:

$$\frac{d \ln P}{dn_i} + \alpha + \beta \varepsilon_i = 0$$

and this is true for all values of i.

To get a value for ln P we use Stirling's approximation

$$\ln x! = x \ln x - x \,.$$

In this the error is proportional to $1/x$ and is thus negligible when we are dealing with perhaps 10^{20} particles.

Now $P = cW$ where c is a simple proportionality constant

$$\therefore \ln P = \ln c + \ln N! - \sum_i \ln n_i! \,,$$

and when we use Stirling's approximation

$$\ln P = \ln c + \ln N! - \sum_i (n_i \ln n_i - n_i)$$

provided that all the energy levels contain a sufficiently large number of particles for Stirling's approximation to be valid.
Thus,

$$\frac{d(\ln P)}{dn_i} = -\ln n_i$$

substituting

$$\ln n_i = \alpha + \beta \varepsilon_i$$

or

$$n_i = e^{\alpha} e^{\beta \varepsilon_i} \,.$$

We have now deduced the basic distribution equation, which tells us how many particles (n_i) occupy any particular energy state (ε_i). Of course, the two constants α and β remain to be evaluated, but before doing this some discussion of the distribution equation is merited.

The assumption that all the levels contain a large number of particles will not always be justified and it is then necessary to modify the derivation. Instead of considering the levels individually we consider them in bundles, so chosen that each bundle contains the required large number of particles but only encompasses a very narrow energy range. If there are p_i levels included in the ith bundle, then the number of ways in which the n_i particles can be distributed within the bundle is $p_i^{n_i}$ (for example, in the table on p. 14, if the bundle consists of just two levels, the last column confirms that the number of ways is 2^{n_i}). This factor must be included in our modified equation for W, which now becomes

$$W = \frac{N!}{n_1! n_2! n_3! \ldots} \times p_1^{n_1} p_2^{n_2} p_3^{n_3} \ldots \tag{2.5}$$

When this expression for W is used in the rest of the derivation above, then the number of particles in any level, n_i, becomes

$$n_i = p_i \mathrm{e}^\alpha \mathrm{e}^{\beta \varepsilon_i}$$

which is the *Boltzmann Distribution Law*. In this form the distribution law applies quite generally. If the energy spread of the levels in a bundle tends to zero, so that there are now p_i levels of energy ε_i the ith level is said to have a 'degeneracy' of p_i.

We note that this equation tells us about the most probable behaviour of a large number of particles, but tells us nothing about less probable behaviour or small numbers of particles. However, as the number of particles increases the most probable distribution increases in importance relative to other distributions roughly as $N^{\frac{1}{2}}$, so that for a normal chemical assemblage we need only consider the most probable case.

Fermi-Dirac Statistics

We now consider the first of our quantum-mechanical cases, in which the particles are indistinguishable and the total wave-function is antisymmetric with respect to the interchange of identical particles.

Again our system contains a large number, N, of these indistinguishable particles whose statistical behaviour we are about to determine. The resulting statistics are known as 'Fermi-Dirac statistics' and the particles as 'fermions'. The difference between particles which obey Boltzmann statistics, in which no symmetry condition is imposed, and those which obey Fermi-Dirac statistics can be illustrated as follows. Consider two wave-functions, Ψ_A and Ψ_B, and two particles, (1) and (2). Now we can associate particle (1) with wave function Ψ_A and particle (2) with Ψ_B. Thus the total wave-function is

$$\Psi_{total} = \Psi_A(1)\Psi_B(2)\,.$$

This wave-function has no symmetry with respect to interchange of the particles (1) and (2) and we are dealing with the Bolzmann case. In order to make Ψ_{total} antisymmetric we write it as the linear combination

$$\Psi_{total} = \Psi_A(1)\Psi_B(2) - \Psi_A(2)\Psi_B(1)\,.$$

In this equation, when 1 is written for 2 and 2 is written for 1, Ψ_{total} becomes $-\Psi_{total}$ as required for a change of sign on exchanging particles.

The consequences of an antisymmetric wave-function on the properties of the assembly of particles are profound and we shall consider one example of a particle which obeys Fermi-Dirac statistics: the electron. Electrons in atoms can be described by total wave-functions which must include both the spatial configuration and the spin of the electron. The latter can have either of two values, which we will call (α) and (β). Thus we can further specify Ψ_A and Ψ_B, which are now atomic wave functions, by the spin of the electrons they contain. The possibilities are $\Psi_A^\alpha, \Psi_A^\beta, \Psi_B^\alpha$ and Ψ_B^β. Now we can write one acceptable distribution of electrons as

$$\Psi_{total} = \Psi_A^\alpha(1)\Psi_B^\alpha(2) - \Psi_A^\alpha(2)\Psi_B^\alpha(1)\,.$$

In this equation the electrons have parallel spins—both are α—and provided that $\Psi_A \neq \Psi_B$ we have maintained our requirement that Ψ_{total} be antisymmetric. Note however that if $\Psi_A = \Psi_B$ the total wave-function vanishes, i.e. two electrons with parallel spins must be in different atomic orbitals.

An alternative and acceptable way of writing Ψ_{total} is to allow one of the electron spins to be reversed. A possible distribution of electrons is

$$\Psi_{total} = \Psi_A^\alpha(1)\Psi_B^\beta(2) - \Psi_A^\alpha(2)\Psi_B^\beta(1)\,.$$

As the equation is written the electrons are in different orbitals and have different spins. However, in this case we can allow Ψ_A and Ψ_B to become the same orbital, Ψ. Then

$$\Psi_{\text{total}} = \Psi^\alpha(1)\Psi^\beta(2) - \Psi^\alpha(2)\Psi^\beta(1)$$

which is again antisymmetric when electrons 1 and 2 are interchanged and vanishes if the electrons have the same spin, i.e. if $\alpha = \beta$. We have now reached the important conclusion that two electrons in the same orbital cannot have the same spin, a result that will clearly affect the distribution of electrons among the energy levels of an atom. It requires the three quantum numbers n, l, and m_l to define an atomic orbital (due to there being three cartesian or polar space co-ordinates) and a fourth quantum number m_s to specify the electron spin, so that we may conclude that 'No two electrons in an atom can have all four quantum numbers the same'. This is of course the more familiar form of the *Pauli Exclusion Principle*.

The general conclusion of our discussion is that there is a restriction on the number of Fermi-Dirac particles which can occupy an energy level to one of each spin (α or β) per level. This restriction necessarily leads to a distribution different from the Boltzmann distribution. To illustrate the effect of the restriction we consider in detail some cases of small numbers of particles with the same spin in a system of energy levels (see next page). This time we shall from the start allow p_i levels to have a particular energy, ε_i (i.e. a degeneracy of p_i). n_i must, of course, not exceed p_i (since there can only be one particle per level) and each time we increase p_i we can build the distribution for a particular n_i in the table from previous entries. For example, take the case of putting 2 particles ($n_i = 2$) in a four-fold degenerate set of levels ($p_i = 4$). If we leave the first level empty, we are then putting 2 particles in 3 levels which the previous entry for $n_i = 2$ shows can be done in 3 ways. If we put a particle in the first level we are left with the task of putting one particle in a choice of three levels (or 'boxes') which can also be done in 3 ways. The total number of ways is therefore $3+3 = 6$. The general conclusion is that the number of ways in which n_i indistinguishable particles can be distributed in p_i levels, each of which can accommodate one particle, is

$$W_{n_i,p_i} = \frac{p_i!}{n_i!(p_i - n_i)!}.$$

p_i (number of levels with same energy)	n_i (number of particles)	W_{n_i, p_i} (number of distributions)
2	0	1
	1	2 (i.e. the particles can be in either level)
	2	1(Here we have to have one particle in each level)
3	0	1
	1	$2 + 1 = 3$
	2	$1 + 2 = 3$
	3	1
4	0	1
	1	$3 + 1 = 4$
	2	$3 + 3 = 6$
	3	$1 + 3 = 4$
	4	1

When we consider the entire system the total number of ways in which a given distribution can occur is

$$W = \prod_i W_{n_i, P_i}$$

where Π_i is the product for all values of i of the various distributions. With this modified probability the analysis follows the Boltzmann case and a new formula for the most probable distribution is derived as:

$$n_i = p_i / (e^{-\alpha} e^{-\beta \varepsilon_i} + 1).$$

This formula applies to all particles with antisymmetric wave-functions. Experience has shown that particles with half-integral spin obey these statistics, e.g. electrons, protons and ^{3}He nuclei.

Bose-Einstein Statistics

The second quantum-mechanical case we shall consider is a system of N indistinguishable particles which have wave-functions that are symmetrical with respect to interchange of particles. This time there is no limit to the number of particles that we can put in a level because changing any pair will not change the symmetrical wave-function.

The distribution of a small number of particles will now be of the following form, starting with our table which we shall then generalize:

p_i(number of levels of same energy)	n_i(number of particles)	W_{n_i,p_i} (number of indistinguishable distributions)
2	0	1
	1	2
	2	3
	3	4
	4	5
3	0	1
	1	$1 + 2 = 3$
	2	$1 + 2 + 3 = 6$
	3	$1 + 2 + 3 + 4 = 10$

Again we can build up a table from previous entries. Thus if we consider the case in which we have to place three particles in three boxes, A, B and C, without any restrictions, we can add up the possibilities as follows:

$$3 \text{ in A leaving 0 in B and C} = 1 \text{ way}$$
$$2 \text{ in A leaving 1 in B and C} = 2 \text{ ways}$$
$$1 \text{ in A leaving 2 in B and C} = 3 \text{ ways}$$
$$0 \text{ in A leaving 3 in B and C} = 4 \text{ ways}$$

$$\text{Total} = \overline{10} \text{ ways}.$$

The general formula for distributing n_i indistinguishable particles among p_i levels, without any restrictions, is

$$W_{n_i,p_i} = \frac{(n_i + p_i - 1)!}{n_i!(p_i - 1)!} .$$

The total number of ways in which a given distribution can occur is

$$W = \prod_i W_{n_i,P_i} .$$

This time the mathematical analysis, which follows the Boltzmann case after obtaining W, yields the formula

$$n_i = \frac{p_i}{e^{-\alpha}e^{-\beta\varepsilon_i} - 1}$$

for the most probable distribution of a particular energy and number of particles.

Experience has shown that particles with integral spin (including zero) have symmetric wave functions and obey Bose-Einstein statistics. These are known as 'bosons', examples being deuterons, photons, and ^{4}He atoms.

Chemical Statistics

For almost all systems of chemical interest $e^{-\alpha}e^{-\beta\varepsilon_i}$ is very large in comparison with 1. The term $+1$ or -1 which occur in the denominators of the Fermi-Dirac and Bose-Einstein formulae can therefore be neglected. Exceptions of practical importance occur when we are dealing with electrons in metals or where we are close to the absolute zero.

When the term $+1$ or -1 can be neglected, both the Fermi-Dirac and Bose-Einstein formula reduce to the classical Boltzmann formula,

$$n_i = p_i e^{\alpha} e^{\beta\varepsilon_i} .$$

Under most circumstances, therefore, systems of chemical interest follow the Boltmann distribution law and the distribution of particles among energy levels is not significantly affected by the symmetry properties of the wave-functions which described the system.

We now go on to derive values for the constants α and β which occur in the distribution law, and thus to relate the statistical to the thermodynamic properties of the system.

Chapter 3

DISTRIBUTIONS AND THERMODYNAMICS

So far we have determined the ways in which particles distribute themselves among the available energy levels and it is now necessary to make the all-important link between these statistical properties of the assembly and its bulk thermodynamic properties. To do this we shall need to determine the two constants α and β which occur in the Boltzmann equation

$$n_i = p_i e^{\alpha} e^{\beta \varepsilon_i} \,.$$

Once we have determined α and β it will become possible to express many of the bulk properties of a system, for example its internal energy or entropy, in terms of the distribution law. When we do this we shall be looking at these functions in a new way, quite different from the classical approach to the first and second laws of thermodynamics which we discussed briefly in Chapter 1. The connection between the distributions of particles and the thermodynamic properties of the system can be made by a consideration of entropy.

Entropy and Distribution

When the temperature is raised the entropy of any system tends to increase for two reasons; first because the increasing thermal motion of the molecules tends to make the assembly occupy a volume greater than that required to pack the molecules together in the solid state, second because increase in temperatures leads to the promotion of particles from the ground state and low-lying energy states to higher levels. The extent to which this occurs is described by the appropriate distribution law. We should therefore expect a definite relation to exist between entropy, thought of as the 'randomness' of a system, and the probability (P) or number of different ways (W) in which a system with a given number of particles and of constant energy can be realized (remember $P \propto W$). The greater the probability, the greater the entropy. The expected relation does exist and can be rationalized as follows. The entropies of two independent parts of an assembly may be added to give the entropy of the whole system,

$$S = S_1 + S_2.$$

On the other hand if the number of ways of distributing particles in one part is W_1, and the number of ways of distributing particles in the other part is W_2, then the number of ways of distributing the whole system is the product

$$W = W_1 \times W_2,$$

because for any individual distribution of one system all possible distributions of the other are possible. We thus have the different behaviour of entropy which is additive and the distributions which are multiplicative. Since we think of entropy as a measure of randomness, this strongly suggests a relation of the type:

$$S \propto \ln W.$$

With proportionality constant, k

$$S = k \ln W = k \ln(W_1 \times W_2) = k \ln W_1 + k \ln W_2 = S_1 + S_2.$$

Having made the connection between a thermodynamic property (S) and the distribution (W), we must now identify the constants α and β.

The Constants α and β in the Distribution Law

In discussing the distribution of particles among energy levels we are concerned with the extent to which the particles escape from the lowest of the energy levels available to them. It is in this level that they congregate at low temperatures. The zero of energy is chosen to be the lowest level to which the particles are able to fall, and any other, lower, levels which cannot be reached physically are not included. (In this connection it is instructive to realize that theoretically the lowest energy state of some hydrogen in a cylinder would be ^{57}Fe if all the nuclear particles could fuse and the excess mass be released as energy. Fortunately it cannot happen under terrestrial conditions.)

The two arbitrary constants α and β were introduced in the derivation of the distribution law in order to combine the three equations (2.2, 2.3 and 2.4) which we wished to apply to the system. Thus, if in the equation

$$n_i = p_i e^\alpha e^{\beta \varepsilon_i}$$

we set ε_i equal to zero, n_i becomes the number of particles in the lowest level and for non-degenerate levels, where $p_0 = 1$,

$$e^\alpha = n_0 \, .$$

Even if this lowest level were to be degenerate, e^α is just a number, proportional to n_0.

We can use this fact to show what the value of the other constant, β, must be. As we saw above

$$S = k \ln W \, .$$

Now using the results of the last chapter for a non-degenerate system of distinguishable particles and repeating some lines of algebra for the sake of clarity:

$$W = N!/n_1!n_2!n_3! \ldots$$
$$S = k \ln N! - k \sum_i \ln n_i!$$

which on using Stirling's approximation becomes

$$S = k \left\{ N \ln N - N - \sum_i (n_i \ln n_i - n_i) \right\} .$$

This, on using the distribution law for $\ln n_i$, yields

$$S = k \{ N \ln N - \sum_i n_i (\alpha + \beta \varepsilon_i) \}$$

but

$$\sum_i n_i = N \quad \text{and} \quad \sum_i \varepsilon_i n_i = E$$

so that

$$S = k(N \ln N - \alpha N - \beta E). \tag{3.1}$$

We may identify the energy E with the thermodynamic internal energy U and from classical thermodynamics (Chapter 1) we have

$$\left(\frac{\partial S}{\partial U}\right)_V = \frac{1}{T}.$$

Also from Eq. (3.1) remembering that α is a number

$$\left(\frac{\partial S}{\partial U}\right)_V = -k\beta.$$

Therefore

$$\beta = -\frac{1}{kT}.$$

This value of β applies generally, so that the distribution law becomes

$$n_i = n_0 p_i \exp(-\varepsilon_i/kT).$$

This equation describes the relation of the number of particles in a particular level, n_i, to the number of particles in the ground level, n_0. It is also of interest to know the relation between the number of particles in a particular level and the total number of particles, N. This relation can be derived from the distribution law using the fact that the total number is the sum of the numbers of particles in all the individual levels, i.e.

$$N = \sum_i n_i = n_0 \sum_i p_i \exp(-\varepsilon_i/kT)$$

but

$$n_0 = \frac{n_i}{p_i \exp(-\varepsilon_i/kT)}.$$

Therefore,

$$n_i = \frac{N p_i \exp(-\varepsilon_i/kT)}{\sum_i p_i \cdot \exp(-\varepsilon_i/kT)}. \tag{3.2}$$

This equation gives us some useful additional information about the distribution of the particles in their energy levels. It also introduces the dimensionless quantity $\sum_i p_i \exp(-\varepsilon_i/kT)$ which plays a very important role in statistical thermodynamics and is called the '*partition function*' q.

The Partition Functions q and Q

Although Eq. (3.2) gives the mathematical definition of q, it is of interest to enquire into the physical significance of a partition function. In so doing we note first that q is determined by the characteristic features, p_i and ε_i, of a particular set of energy levels. These levels form an energy-level 'ladder', up which a particle may climb. Secondly, we note that the formula for q does not include the number of particles under consideration. It may be looked on as giving information about the way, on average, a particle will be distributed on the 'rungs' of the ladder. When we come to consider an assemblage of N particles, each of which has its own partition function, q, the total partition function of the assembly, Q, is the product of the individual q's. Thus

$$Q = (q)^N .$$

Q is called the 'molar partition function' and q is called the 'molecular partition function'.

We should remember here that at the moment we are dealing with distinguishable, i.e. localized, particles. For a system of non-localized particles a gas, for example, the equation is modified from that for indistinguishable particles:

$$Q = q^N/N!$$

This equation is discussed further in the section on the Sackur-Tetrode equations (p. 38).

It will often happen that a system of particles has more than one series of energy levels available. For example, at room temperature diatomic molecules can have translational, rotational, and vibrational energy. To a good approximation, however, the different energies are independent of one another and each energy-level ladder can be treated separately. In general, when the system is at a very low temperature, $\varepsilon_i \gg kT$ and the number of molecules in any exicited state n_i is:

$$n_i = n_0 p_i \exp(-\varepsilon_i/kT) \sim 0$$

so that all the particles are in the bottom level. As the system is warmed, at constant volume, some of the particles acquire sufficient energy to populate higher levels and n_i becomes non-zero. The total number of particles, N, is the sum of the numbers in the individual levels, including the bottom level: i.e.

$$N = \sum_i n_i = n_0 \sum_i p_i \exp(-\varepsilon_i/kT) = n_0 q$$

and

$$q = \frac{N}{n_0}.$$

This simple equation shows that for a closed system, in which N is constant, the partition function is a measure of the extent to which the particles are able to escape from the ground state. It is also clear from this equation that q is a pure number, which can range from unity (not zero) at the absolute zero (when $n_0 = N$ and only one level is accessible), to an indefinitely large number as the temperature is increased because fewer and fewer particles are left in the ground state, and an indefinitely large number of states is accessible (e.g. for a gas at one atmosphere and at room temperature $q_{\text{trans}} \sim 10^{25}$ where q_{trans} is the molecular translational partition function). We can now see that if the energy-level ladder has many closely spaced states the particles will find it is easy to leave the ground state and q will rise very rapidly as the temperature of the system is raised. Conversely, a few widely spaced levels will lead to a small q. An example of the former state of affairs is the translational motion of a gas, whereas the latter is typical of the vibrations of a light diatomic gas at room temperature.

We can illustrate further the effect of temperature on the populations of energy levels by considering a particular case. Let us take as an example a series of non-degenerate, equally spaced levels. (These are the energy levels of a simple harmonic oscillator to which the lower vibrational levels of a diatomic molecule are a reasonable approximation.) If the levels are numbered 0, 1, 2, 3...i and the spacing between adjacent levels is ε the energies of the levels are 0, ε, 2ε, 3ε...$i\varepsilon$. The fraction of molecules in any energy level, n_i/N can then be calculated from Eq. (3.2). In general we notice that when ε is small compared with kT many upper levels are significantly populated but that if we change to a system in which the spacing of the energy levels is greater, the occupancy of the upper levels, at constant temperature, drops rapidly. The results of calculations of the fractional occupation of the levels for a range of values of ε is shown in Fig. 3.1. To give

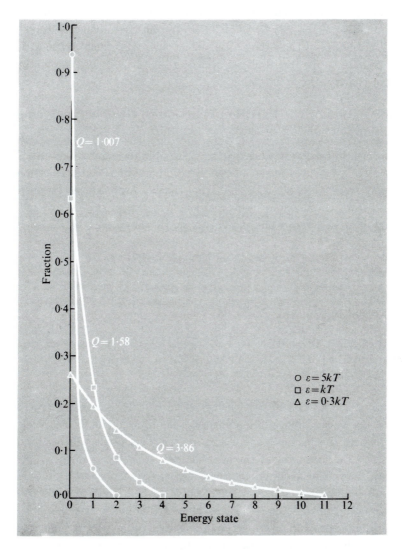

Fig. 3.1. Fractional occupation of levels for various energy-level separations.

an idea of some molecules for which the distributions approximate to those shown in Fig. 3.1 we note that at room temperature, $kT \approx 200$ cm^{-1}. Then for the vibration of carbon monoxide (frequency = 2170 cm^{-1}) at about 620 K, $\varepsilon \approx 5\,kT$; for iodine (frequency = 214 cm^{-1}) at room temperature

$\varepsilon \approx kT$; and for molecules of Na_2 (frequency $= 159$ cm^{-1}), which occur in sodium vapour at 750 K, $\varepsilon \approx 0.3\ kT$. We must remember however that the distributions of these molecules among their vibrational levels will only approximate to the distributions drawn in Fig. 3.1 because the vibrations are not exactly simple harmonic.

The Relation of the Partition Function to the Thermodynamic Functions

We have just seen that when particles are able to escape easily from the zeroth level, this gives rise to a large partition function. But the occupation of many energy levels is also the signal that an aspect of a system has a large entropy. It therefore seems reasonable to expect that a close relation should exist between S and q. This is so. If we go back to Eq. (3.1) we have

$$S = k(N \ln N - \alpha N - \beta U)$$

but

$$e^\alpha = n_0 = \frac{N}{q}, \alpha = \ln N - \ln q$$

$$\beta = -\frac{1}{kT}.$$

Thus

$$S = k \ln q^N + \frac{U}{T}.$$

This equation applies to particles which are localized, such as crystalline solids. As we saw above, a modification has to be made for cases such as gases where the particles are free to move. This point will be discussed further at the beginning of the next chapter.

The thermally acquired internal energy, U, of a system at constant volume is often associated with molecular motion, such as translation or rotation, and in these cases a high value of U implies that the molecules are travelling fast or rotating rapidly. Put another way, a high value of U means that many translational or rotational energy levels are occupied. This in turn suggests that there should be a functional relationship between U and q, which is indeed the case. It can be derived as follows. By definition

$$U = \sum_i \varepsilon_i n_i = N \sum_i \frac{\varepsilon_i p_i \exp(-\varepsilon_i/kT)}{q}.$$

In order to carry out the summation we note that

$$\frac{\partial}{\partial T}(p_i \exp(-\varepsilon_i/kT)) = \frac{\varepsilon_i p_i \exp(-\varepsilon_i/kT)}{kT^2}$$

and substituting the value for $\varepsilon_i p_i \exp(-\varepsilon_i/kT)$ from this last equation we arrive at

$$U = \frac{\sum_i NkT^2(\partial/\partial T)(p_i \exp(-\varepsilon_i/kT))}{q}$$

but, again by definition

$$q = \sum_i p_i \exp(-\varepsilon_i/kT)\,;$$

therefore

$$U = \frac{NkT^2}{q}\frac{\partial q}{\partial T} = NkT^2\frac{\partial(\ln q)}{\partial T}$$

or

$$U = kT^2\frac{\partial(\ln Q)}{\partial T}$$

since

$$Q = q^N\,.$$

Now that we have related both internal energy, U, and entropy, S to the partition function, q, it is an easy task to produce equations relating all other thermodynamic functions and properties to q, using the classical thermodynamic formulae quoted in Chapter 1.

We have now found the relation between distributions and thermodynamics which we sought. All we need in order to calculate the thermodynamic functions are the appropriate partition functions. Their calculation is the subject of our next chapter.

Chapter 4

DETERMINATION OF PARTITION FUNCTIONS

As we have seen in the last chapter, a knowledge of the partition function is extremely useful in statistical thermodynamic calculations. The partition functions contains all the information about equilibrium states; if we know q (and its temperature dependence) then we can calculate all the thermodynamic functions. For idealized and simple systems, where the energy levels are known accurately and where we can reasonably assume that the energy of the system is simply the sum of the energies of the individual particles, this calculation is straightforward. Although real systems depart to some extent from ideal behaviour, we can often deduce formulae for partition functions which are very good approximations to reality even when we use an idealized model.

When doing the calculation, however, we need to take into account the fact that the number of distributions, W, available to a constant number of particles depends on whether the particles are localized or non-localized.

Localized and Non-Localized Systems

As localized and non-localized systems we can take as examples a mole of a monatomic substance which can exist in the extremes of an ideal crystalline solid or an ideal gas. The molecules are identical in both cases but there are considerable differences between the two situations.

In the solid we may have a molecule with energy a at one lattice point X and another with energy b at a different point Y. This situation is physically distinguishable from the case where the energy of the molecule at X is b and that at Y is a. Thus both 'states' go to make up the sum of distinguishable state W. On the other hand if we have two gas molecules with translational energies a and b respectively the situation is different. At one instant the molecule with energy a is at point X in space and that with energy b is at point Y. Since all positions of particles are possible in the gas, at some other time the molecule with energy b may be at X and that with a at Y. This second situation is indistinguishable from the first. Consequently we can only count such a state as one contribution to the number of distributions W.

Let us see the effect of this difference on the value of W by considering first a gas consisting of three particles with different energies a, b and c contained in a box. There is only one possible distribution for the three particles since their positions cannot be defined. Now let us consider the same three particles in a lattice which we may consider as three indentical boxes distinguishable by their position in spaces. We could label the boxes 1, 2 and 3

$$\begin{array}{ccc} \square & \square & \square \\ 1 & 2 & 3 \end{array}$$

The following distributions are now possible

$$\begin{array}{ccc} \boxed{a} & \boxed{b} & \boxed{c} \\ 1 & 2 & 3 \end{array}$$

$$\begin{array}{ccc} \boxed{a} & \boxed{c} & \boxed{b} \\ 1 & 2 & 3 \end{array}$$

$$\begin{array}{ccc} \boxed{b} & \boxed{a} & \boxed{c} \\ 1 & 2 & 3 \end{array}$$

$$\begin{array}{ccc} \boxed{b} & \boxed{c} & \boxed{a} \\ 1 & 2 & 3 \end{array}$$

$$\begin{array}{ccc} \boxed{c} & \boxed{a} & \boxed{b} \\ 1 & 2 & 3 \end{array}$$

$$\begin{array}{ccc} \boxed{c} & \boxed{b} & \boxed{a} \\ 1 & 2 & 3 \end{array}$$

i.e. there are 6=3! distributions.

If we had a fourth particle with energy d and four boxes then the number of possible distributions would be many more. When we put particle a in box 1 there are 6 ways of putting the other three in the remaining boxes, similarly if b is in box 1 and so on. Thus the number of possible distributions is now $4 \times 6 = 4 \times 3! = 4!$ Generalizing this result; for N particles, if there was one distribution in the ideal-gas situation there will be $N!$ possibilities in the ideal crystal. The entropy is thus changed and we deduce that:

$$S_{\text{solid}} - S_{\text{gas}} = k \ln(W_{\text{solid}}/W_{\text{gas}}).$$

In general:

$$S_{\text{localized}} - S_{\text{non-localized}} = k \ln N! = Nk(\ln N - 1)$$

(using Stirling's approximation).

It should be stressed that we are dealing in both cases with intrinsically indistinguishable particles, and that the entropy difference is not a consequence of quantum effects. It merely follows because, in a solid, even if the particles are exactly the same, the lattice sites are distinguishable on the basis of their unique coordinates. But when we consider a gaseous non-localized assembly we must remember that all the molecules in a particular level or bundle of levels (p) are indistinguishable and permutations among these particles do not lead to new states. This consideration reduces the number of ways of realizing one particular state, and so W and S are also reduced, but leaves the relative populations of the levels unaffected.

Separation of the Partition Function

If we remember that, physically, the partition function tells us how molecules are distributed among the available energy levels and that there are several sorts of energy, we can separate various contributions to the partition function. We can often express the total energy of a molecule as the sum of the translational, rotational, vibrational and electronic energy terms

$$E = E_{\text{trans}} + E_{\text{rot}} + E_{\text{vib}} + E_{\text{el}}.$$

The partition functions (being measures of probabilities) will then be a produce of corresponding terms

$$q = q_{\text{trans}} \times q_{\text{rot}} \times q_{\text{vib}} \times q_{\text{el}}$$

each of which may be considered separately. A further simplification arises from the formulae derived in the last chapter in which, usually, the thermodynamic expressions (e.g. for U and S) involve $\ln q$, so that:

$$\ln q = \ln q_{trans} + \ln q_{rot} + \ln q_{vib} + \ln q_{el}.$$

To calculate any partition function we start with its mathematical definition

$$q = \sum_i p_i \exp -(\varepsilon_i/kT).$$

Hence for the calculation we need to know only the energy levels, their degeneracies, and the temperature. Each of the above partition functions will now be discussed in turn.

Translational Partition Function

For a monatomic gas such as helium the translational component is the only partition function greater than unity under normal conditions. We commence by considering a particle in a one dimensional box. This particle must fulfil the de Broglie relation:

$$mv = \frac{h}{\lambda}.$$

For our particle the wavelength λ is controlled by the length of the box l. And integral number of half wavelengths must fit into the box (Fig. 4.1).
Thus:

$$\lambda = \frac{2l}{n}$$

and

$$mv = \frac{nh}{2l}.$$

The kinetic energy of a molecule is then given by

$$\text{K.E.} = \frac{1}{2}mv^2 = \frac{n^2 h^2}{8ml^2}.$$

Hence for a particle moving in one direction

$$(q_{trans})_{1D} = \sum_{n=1}^{n=\infty} = \exp(-n^2 h^2/8ml^2 kT).$$

Since the energy levels are very close together we can replace the summation by an integral without introducing any serious mathematical errors. Note, however, that this does not imply that the energy of the levels becomes continuous, quantization remains important because Planck's constant remains in the expression. Making this approximation, and the equally trivial one of changing the lower limit of integration from 1 to 0, we have, for one-dimensional translation,

$$(q_{\text{trans}})_{1D} = \int_0^\infty \exp(-n^2 h^2 / 8\, m l^2 kT)\mathrm{d}n\,.$$

If we let

$$x^2 = \frac{n^2 h^2}{8\, m l^2 kT}$$

then

$$(q_{\text{trans}})_{1D} = \frac{1}{h}(8\, mkT)^{\frac{1}{2}} \int_0^\infty \mathrm{e}^{-x^2}\,\mathrm{d}x\,.$$

Now $\int_0^\infty \mathrm{e}^{-x^2}\,\mathrm{d}x$ is a standard integral having the value $\frac{1}{2}\sqrt{(\pi)}$. Thus

$$(q_{\text{trans}})_{1D} = \frac{(2\pi\, mkT)^{\frac{1}{2}} l}{h}\,.$$

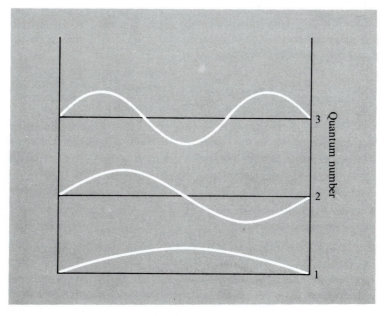

Fig. 4.1. Wave-functions and energy levels of a particle in a one-dimensional box.

For each of the three possible directions in space we obtain a similar expression except that l is replaced by dimensions of the 'box' in these other directions. For three dimensions the translational partition function is the product of these terms, i.e.

$$q_{\text{trans}} = (2\pi \, mkT/h^2)^{\frac{3}{2}} V \,.$$

Here V is the volume. The translational partition function unavoidably involves the volume, since only by relating the motion of the particle to the dimensions of the container in which it moves can any formulation of discrete energy levels be reached.

The Sackur-Tetrode Equation

Under normal conditions an ideal monatomic gas has translational energy only and hence its complete partition function is simply the translational function. Remembering our discussion above of localized and non-localized systems, we must take account of the indistinguishability of gas molecules. To do this, using the result from the last section (but now for one mole of gas rather than for the single particle we discussed above) we write

$$Q_{\text{trans}} = \frac{1}{N!} (q_{\text{trans}})^N$$

Here the factor $1/N!$ arises because the particles are not localized and the individual qs for each molecule are multiplied since they are, in a sense, probabilities. A simple view, which is true when we have an infinite number of states, is that if q_A is the partition function of particle A and q_B the partition function of the identical particle B, then $q_{A,B} = q_A \times q_B$ (where $q_{A,B}$ is the partition function of the system which includes both A and B). Hence for N particles (one mole) $Q = q^N$ if the particles are localized, but $q^N/N!$ if they are non-localized. Thus

$$Q_{\text{trans}} = (1/N!)\{(2\pi \, mkT/h^2)^{\frac{3}{2}} V\}^N \,. \tag{4.1}$$

The internal energy of one mole of gas is given by the formula derived in the last chapter:

$$U = NkT^2 \frac{\partial(\ln Q)}{\partial T} \,.$$

Substituting the value of Q_{trans} from Eq. (4.1) we obtain the value of the energy associated with translational motion:

$$U = NkT^2 \cdot \frac{3}{2} \cdot \frac{1}{T} = \frac{3}{2} NkT.$$

This is also the result of the simple kinetic theory of gases, if the proportionality constant k is the Boltzmann constant.

We can now proceed to derive an expression for the entropy of the ideal monatomic gas. It is convenient to use the Stirling approximation in the form

$$N! \approx (N/e)^N$$

(on taking logarithms this yields the more common form $\ln N! = N \ln N - N$) so that Eq. (4.1) becomes

$$Q = \left[\frac{(2\pi\, mkT)^{\frac{3}{2}} e\, V}{Nh^3} \right]^N$$

i.e.

$$\ln Q = N \ln \left[\frac{eV}{Nh^3} (2\pi\, mkT)^{\frac{3}{2}} \right].$$

Using the result of Chapter 3

$$S = k \ln Q + \frac{U}{T}$$

which was deduced for an assemblage of localized particles, we can substitute to obtain S_{molar}. In making this substitution we do not need to take further account of the fact that we are dealing with a gas rather that a solid, since the factor $1/N!$ in Eq. (4.1) has already covered the difference between the two systems.

The molar entropy is thus

$$S_{\text{molar}} = \frac{3}{2} R + R \ln \frac{eV}{Nh^3} (2\pi\, mkT)^{\frac{5}{2}}$$

$$= R \ln \left(\frac{2\pi\, mkT}{h^2} \right)^{\frac{1}{2}} \cdot \frac{V\, e^{\frac{5}{2}}}{N}$$

which is the *Sackur-Tetrode* equation.

An alternative form of this equation may be obtained by removing the volume term V by using the ideal-gas equation (for 1 mole)$PV = RT = NkT$ i.e.

$$S_{\text{molar}} = R \ln \left(\frac{2\pi \, mkT}{h^2} \right)^{\frac{1}{2}} \cdot \frac{kT}{P} \cdot e^{\frac{5}{2}}.$$

The formulae for the translational contributions to thermodynamic functions are summarized in Appendix 1 including the forms most appropriate for numerical applications.

Rotational Partition Function

Diatomic Molecules

As a first approximation the diatomic molecule may be considered as a rigid rotating dumbbell (Fig. 4.2) whose energy is $\frac{1}{2}I\omega^2$ where I is the moment of inertia and ω the angular velocity. Quantum theory tells us that angular momentum (in this case $I\omega$) is quantized in units of $h/2\pi$, *viz.*

$$I\omega = \sqrt{\{J(J+1)\}}h/2\pi\,.$$

Thus the rotational energy, $\frac{1}{2}I\omega^2$ is given by

$$E_{\text{rot}} = \frac{J(J+1)h^2}{8\pi^2 I}$$

or more concisely

$$E_{\text{rot}} = BJ(J+1)$$

where J is a quantum number and $B = h^2/8\pi^2 I$. A more detailed treatment of the rigid rotor shows that each level has a degeneracy of $(2J+1)$ so that in the distribution equation the statistical weight factors p_J will also be $(2J+1)$.

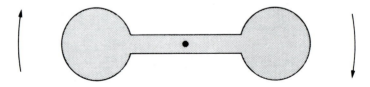

Fig. 4.2. The rigid rotor.

Thus

$$q_{\text{rot}} = \sum_J (2J + 1)\exp(-J(J+1)h^2/8\pi^2 IkT).$$

For most molecules (except hydrogen) the rotational energy levels are close enough together so that, again for mathematical convenience, we can replace the summation by the integration, i.e.

$$
\begin{aligned}
q_{\text{rot}} &= \int_0^\infty (2J+1)\exp(-J(J+1)h^2/8\pi^2 IkT)\mathrm{d}J \\
&= \frac{-8\pi^2 IkT}{h^2} \int_0^\infty \frac{\mathrm{d}}{\mathrm{d}J}\exp\{-J(J+1)h^2/8\pi^2 IkT\}\mathrm{d}J \\
&= \frac{8\pi^2 IkT}{h^2}.
\end{aligned}
$$

The energy separation between two adjacent levels is:

$$
\begin{aligned}
\Delta E &= BJ(J+1) - B(J-1)J \\
&= 2BJ.
\end{aligned}
$$

Because $B \propto 1/I$ the energy levels will be more widely spaced in lighter molecules, and the lighter the molecule the worse the approximation of the summation to an integral. In the case of HCl gas at room temperature, for example, the exact value of q_{rot} obtained by summation, is 20·39 while the value from the analytically integrated formula is 19·98; a small discrepancy even for this comparatively low moment of inertia molecule. However for the isotopes of hydrogen numerical evaluation of q_{rot} is necessary. If we define $x = h^2/8\pi^2 IkT$ then the summation for q_{rot} can be expanded and all properties expressed as series expansions in x. This is done in Appendix 2 together with a summary of useful working expressions.

An additional complication arises with homonuclear diatomics. We shall return to this point in Chapter 8; suffice it to note here that the partition function for symmetric molecules is modified by a symmetry number σ which is 2 for homonuclear diatomics and 1 for heteronuclear diatomics. Thus, in general, for diatomic molecules

$$q_{\text{rot}} = 8\pi^2 IkT/\sigma h^2.$$

Polyatomic Molecules

If the polyatomic molecule is linear, like CO_2 or N_2O, it has only one moment of inertia and the diatomic molecule formula can be used. In general, however, a molecule has three moments of inertia about mutually perpendicular axes, I_A, I_B and I_C. The classical rotational partition function then looks like a product of terms for each axis

$$q_{rot} = 8\pi^2 (8\pi^3 I_A I_B I_C)^{\frac{1}{2}} (kT)^{\frac{3}{2}} / \sigma h^3 .$$

The symmetry number, σ, is the number of indistinguishable positions into which the molecule can be turned by simple rotation and is introduced to avoid counting the same levels twice: for example for H_2O, $\sigma = 2$; for NH_3, $\sigma = 3$; for CH_4, $\sigma = 12$; for C_6H_6, $\sigma = 12$.

Vibrational Partition Function

The solution of the Schrödinger equation for a simple harmonic oscillator gives energy levels which are equally spaced, $E_v = (v + \frac{1}{2})h\nu$, while an anharmonic diatomic molecule has levels which generally get closer together with increasing vibrational quantum number v. Many molecules approximate to the simple-harmonic-oscillator model at low v values but all become increasingly anharmonic at high v values; eventually, of course, they dissociate and are not oscillators at all. The energy levels are illustrated in Fig. 4.3.

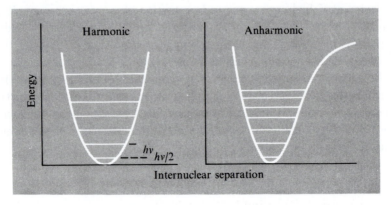

Fig. 4.3. Energy levels of the harmonic and anharmonic oscillators.

It is a reasonable approximation, however, to use the simple-harmonic-oscillator formula at ordinary temperatures since most of the molecules will be in the lowest few energy levels.

Then for each mode of vibration

$$E_{\text{vib}} = (v + \frac{1}{2})h\nu$$

where

$$\nu = \frac{1}{2\pi}\sqrt{\frac{k}{\mu}}$$

(k is the force constant of the bond and μ is the reduced mass, $m_1 m_2/(m_1 + m_2)$ of the two atomic masses m_1 and m_2). Then

$$q_{\text{vib}} = \sum_v \exp(-[v + \frac{1}{2}]h\nu/kT)$$

$$= \exp\left(-\frac{h\nu}{2kT}\right) \sum_v \exp(-vh\nu/kT).$$

The summation is of an infinite geometrical progression. The factor before the summation, $\exp(-h\nu/2kT)$, which represents the zero-point energy, can be ignored if q_{vib} relates only to the population of accessible states. Of these, the state with energy $\frac{1}{2}h\nu$ is the lowest. The geometrical progression can be expanded as

$$1 + e^{-h\nu/kT} + e^{-2h\nu/kT} + e^{-3h\nu/kT} + \cdots$$

with a common ratio of $e^{-h\nu/kT}$. The formula for the sum to infinity of a geometric progression

$$a + ar + ar^2 + ar^3 + \cdots$$

is

$$\frac{a}{1-r} \quad \text{for} \quad r < 1.$$

Therefore

$$q_{\text{vib}} = (1 - e^{-h\nu/kT})^{-1}.$$

If there are several vibrational modes we have to take the product of such terms, one for each vibration.

When substituted into the formulae for the various thermodynamic functions (Chapter 3) it is convenient if $h\nu/kT$ is replaced by a single symbol, say u. In this way tables of vibrational contributions to the thermodynamic functions $U, H, G, S, C_{\text{vib}}$ can be given in terms of u. These expressions and a table of typical results are presented in Appendix 3.

Electronic Partition Functions

$$q_{\text{el}} = \sum_i p_i \exp(-E_i/kT)$$

where E_i are the various electronic energy levels of the molecules. These are generally very widely separated by comparison with, say, vibrational levels as illustrated in Fig. 4.4.

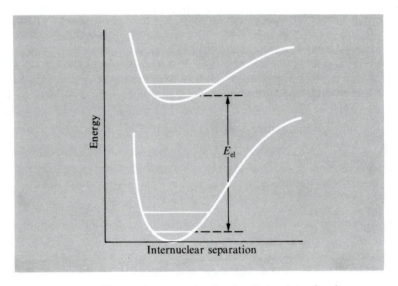

Fig. 4.4. Electronic energy levels of a diatomic molecule.

In this case the electronic partition function is best treated as the explicitly evaluated summation

$$q_{\text{el}} = p_0 \exp(0) + p_1 \exp(-E_1/kT) + p_2 \exp(-E_2/kT) + \cdots$$

Most naturally occurring molecules have ground states with all the electrons paired, two per orbital. In these case p_0 is unity and E_1 the excitation energy of the first excited state is so large that $p_1 \exp(-E_1/kT))$ is negligible and

$$q_{el} = 1.$$

If the ground state has unpaired electrons then the statistical weight will not be unity but is equal to the degeneracy of the lowest level. In a very few molecules there are low-lying excited states. For example, the diatomic species MgO, which can be produced in the gas phase by heating MgO above 1000 K, has a closed-shell ground state, but an excited open-shell state with two parallel electron spins (a triplet) which is so close in energy that it contributes significantly to the partition function.

The Hydrogen Atom

The hydrogen atom provides an amusing problem concerned with electronic partition functions. The energy levels are given to high accuracy by the Bohr formula, or from the solution of the Schrödinger equation for the hydrogen atom.

The energy of level n is (Fig. 2.1)

$$E_n = -R_\infty/n^2$$

where R_∞ is the Rydberg constant and the level has a degeneracy of n^2.

If we measure our excitation energies from the lowest level ($n = 1$) the electronic partition function will be

$$q_{el} = \sum_n n^2 \exp\left\{ -\frac{R_\infty}{kT}\left(1 - \frac{1}{n^2}\right) \right\}.$$

Unfortunately this summation does not converge, owing to the n^2 degeneracy factor. Each term added as n gets larger adds a very small increment, and the sum to $n = \infty$ is infinite, giving the nonsensical answer that the partition function is infinite!

This is a problem that has puzzled many, and some extraordinary solutions have been given. A possible answer is that the Bohr energy formula applies strictly only to an H atom alone in space. In practice, of course, any H atoms which do occur must be in some sort of box, probably a discharge tube or a room or even in a universe which contains galaxies (and other hydrogen atoms). The result of being in a box is that the outer orbitals are perturbed and energies for high n are not as given by the simple formula. These outer obitals are not of any significance in the measured spectrum which is dominated by the inner shell. The 'solution' to the problem of the partition function of hydrogen is then merely that one should recognize that an ideal formula may not give the correct result for non-ideal conditions.

Chapter 5

EQUILIBRIA AND RATES OF REACTION

So far we have discovered how particles are expected to distribute themselves among energy levels and have derived equations for the partition functions for a number of important types of energy-level ladders. Implicit in our discussions has been the assumption that we are dealing throughout with non-reacting systems. However, we can readily extend our treatment to deal with two aspects of chemical reactions of particular importance, namely the final position of the chemical equilibrium and the rate at which the reaction proceeds. This latter application of theories which are essentially to do with equilibrium conditions may seem odd, but as we shall see, some important theories of reaction rates employ the ideas of equilibrium thermodynamics.

Let us start by remembering what it is that a partition function, q, actually tells us. It is a number which gives us the ratio of the total number of molecules present to the number in the lowest level, i.e. q is an indication of the population spread amongst the available levels. Thus ratios of partition functions are also ratios of numbers of molecules and as such can be considered as a type of equilibrium constant.

Equilibrium Constants

We consider first the following profile through a potential surface (Fig. 5.1). From the meaning of the partition functions the equilibrium constant governing the equilibrium between the number of molecules n_A in

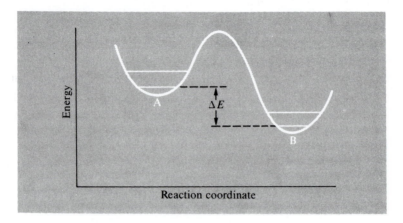

Fig. 5.1. Schematic potential-energy diagram for a simple chemical reaction.

state A and the number n_B in state B should be

$$\frac{n_A}{n_B} = \frac{q_A}{q_B}$$

where q_A and q_B are the respective partition functions. This will, however, be true only if the energies of the two states are both measured from the same level, say the lowest level of state B. If we measure the energy from the lowest level of A for state A molecules and from the lowest level of B for state B molecules we must rewrite the partition function of state A to make allowance for the difference in energy between the two lowest levels. This energy difference is marked as ΔE on the figure and the modified partition function is:

$$q'_A = \sum_i p_i \exp(-[E_{A_i} + \Delta E]/kT)$$

$$= q_A \exp(-\Delta E/kT)$$

$$\frac{n_A}{n_B} = \frac{q_A}{q_B} \cdot \exp(-\Delta E/kT) \, .$$

This is a simple relation between the number of molecules in the two states, their respective partition functions, and the energy gap between the ground states. The ratio n_A/n_B is then the equilibrium constant K_N

expressed in terms of the number of molecules of A and B for

$$A \leftrightharpoons B.$$

We usually express equilibrium constants not in terms of numbers of molecules but in terms of partial pressures or concentrations, and such equilibrium constants are related directly to standard free-energy changes. The necessary modification to the simple expression above is presented in Appendix 4.

The above result can be extended to a more general equilibrium of the type

$$aA + bB \leftrightharpoons cC + dD$$

where

$$K_N = \frac{n_C^c \cdot n_D^d}{n_A^a \cdot n_B^b} = \frac{q_C^c \cdot q_D^d}{q_A^a \cdot q_B^b} \cdot e^{-\Delta E / kT}. \tag{5.1}$$

Again the relation requires minor modification if we consider K_p in terms of partial pressures and this is done in Appendix 4.

The general equation enables us to calculate equilibrium constants directly from a knowledge of energy levels and energy differences (frequently dissociation energies). Alternatively, measurement of equilibrium constants can give us a good estimate of energy differences in favourable cases. These points are illustrated in the following two examples.

The Sodium Atom-Molecule Equilibrium

At temperatures in the region of 1000 K there exists a significant equilibrium between sodium atoms and diatomic sodium molecules

$$2\,Na \leftrightharpoons Na_2.$$

For this reaction the energy difference between the two sides of the reaction is the dissociation energy, D, of the molecules. This may be measured experimentally by first observing the vibrational energy levels of an excited state of Na_2 almost up to the dissociation limit and then subtracting the extra energy of the excited sodium atoms produced by dissociation, knowing the atomic sodium energy levels. The electronic spectrum of Na_2 can be observed by passing white light through heated sodium in a steel tube. The absorption spectrum obtained is illustrated in Fig. 5.2(a).

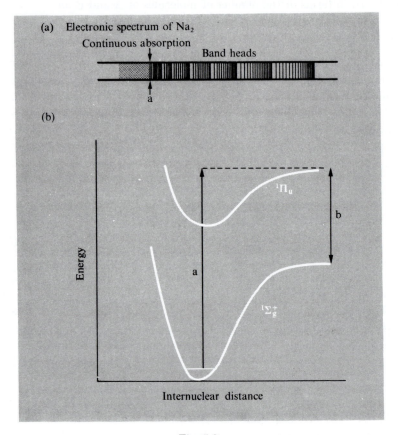

Fig. 5.2.

The onset of continuous absorption corresponds to the energy transition marked a in Fig. 5.2(b). The energy b of this figure corresponds to the excitation of an electron in atomic sodium and is observable as a line in the atomic spectrum of the element. Hence

$$D = a - b.$$

Calculation of the partition function of Na_2 requires a knowledge of the rotational and vibrational levels, both of which are obtained from the electronic spectrum, together with the translational contribution (which demands only a knowledge of molecular weight). For the Na atom the partition function has only translational and electronic contributions. Computation of the equilibrium constant is then merely a question of substituting the appropriate numbers into the formulae. This is done in detail in Appendix 4.

The Dissociation Energy of Fluorine

The above method has been used in reverse in the case of fluorine where a measured equilibrium constant was used to determine $\Delta H^{\ominus}(= D)$ the dissociation energy of the molecule. This calculation is also a typical example of the advantage there may be in using the third law rather than the second law for determinations of enthalpy changes.

For F_2 dissociation the following equilibrium was considered

$$F_2 \rightleftharpoons 2F.$$

This dissociation takes place to an appreciable degree only at elevated temperatures. None the less, two independent series of the difficult high temperature (500–1100 K) vapour-pressure measurements have been made, enabling equilibrium constants (K_p) to be obtained. The integrated form of the Van't Hoff isochore (p. 7)

$$\ln K_p = \frac{-\Delta H^{\ominus}}{RT} + \text{constant of integration}$$

was used and plots of $\ln K_p$ versus $1/T$ gave straight lines with a slope of $-\Delta H^{\ominus}/R$. (The dissociation equilibrium was described either by the ratio $(P_F/P_{F_2}^{\frac{1}{2}})$ as in Fig. 5.3(a) or by (P_F^2/P_{F_2}) as in Fig. 5.3(b). The slope of the latter line is then a factor of two greater than that of the former.) $\Delta H^{\ominus}$ is the desired dissociation energy D. The graphs obtained in the two different experiments are shown in Fig. 5.3. Such a procedure is referred to as a second-law determination of $\Delta H^{\ominus}$ and gives an average value of the heat of dissociation over the range studied. The method suffers from the disadvantage that it depends on the measurement of the slope of a line drawn through points which under high temperature conditions are particularly difficult to get and may have large experimental uncertainties.

If we use a third-law method, involving our statistical thermodynamic formulae, each individual measurement gives an independent value of $\Delta H^{\ominus}$ (instead of using a series of points which together yield one value only of

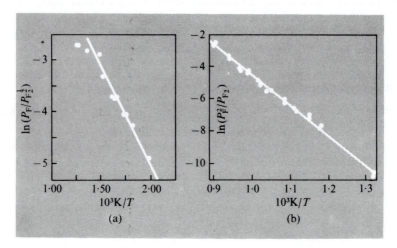

Fig. 5.3. The dissociation of F_2 as a function of temperature.

$\Delta H^{\ominus}$). We can then choose the most convenient temperatures at which to do the experiments and average these more reliable values of $\Delta H^{\ominus}$ to give a great improvement in accuracy. This is done as follows. Since each measurement gives a vapour pressure, it also yields K_p for that temperature. Thus from the second law

$$\Delta G^{\ominus} = -RT \ln K_p = \Delta H^{\ominus} - T\Delta S^{\ominus}.$$

$\Delta G^{\ominus}$ being known, what we need to obtain $\Delta H^{\ominus}$ is a knowledge of $\Delta S^{\ominus}$. $\Delta S^{\ominus}$ can be calculated from the partition functions of the fluorine atoms and molecule, using the formulae of Chapter 3. For this calculation we require the rotational and vibrational energy levels of F_2, which are available from the Raman spectrum, and the masses of the atoms.

The partition functions for the standard states of F_2 and F are given by the same formulae as are used in the above example on Na and Na_2 (Appendix 4). The relevant data to substitute are

$$m_F = 1/2 m_{F_2} = 19 \times 10^{-3}/(6.02 \times 10^{23}) \text{ kg}$$

$$r = 1.44 \times 10^{-10} \text{ m}$$

$$\frac{h\nu_0}{kT} = \frac{6.62 \times 10^{-27} \times 3 \times 10^{10} \times 892}{1.38 \times 10^{-16} \times 10^3}.$$

For the electronic contributions, $p_0(F_2) = 1$, but the atomic ground state is ^2P. The lower, $^2P_{3/2}$, component has degeneracy $p_0(F) = 4$ and the $^2P_{1/2}$ component which lies 404 cm^{-1} higher in energy has $p_1(F) = 2$.

The accuracy of this third law determination is significantly greater than the second law treatment and was highlighted by Barrow and Stamper. They used the partition functions in a third law method to calculate the value of $\Delta H^\ominus$ at the absolute zero (i.e. ΔH_0^0) and were consequently able to show not only the scatter in the measurement of earlier second law attempts but also that there appears to be a systematic drift in the points shown in Fig. 5.3(a). This is shown in Fig. 5.4 from which it was concluded that the best value of ΔH_0^0 is 153.6 ± 0.54 kJ mol^{-1}.

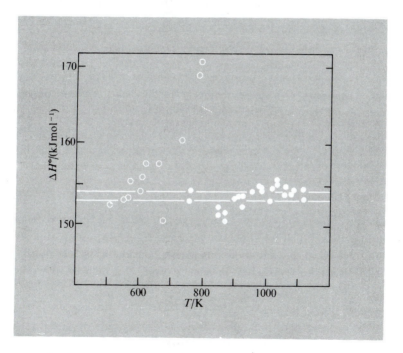

Fig. 5.4. Third-law calculation of dissociation of dissociation energy F_2 using results of Fig. 5.3a (o) and Fig. 5.3b (●).

The measured value of the dissociation energy of fluorine is particularly interesting since it seems to be out of step with the dissociation energies of the other halogens as can be seen in Fig. 5.5.

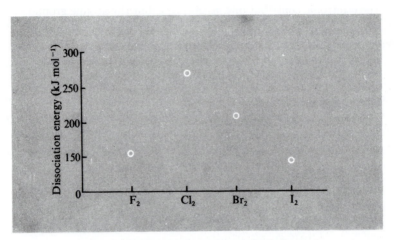

Fig. 5.5. Dissociation energies of the halogens.

The reason for this variation is not completely clear but it has been suggested that part of the anomaly lies in the dissociation energy of chlorine which is bigger than might be expected owing to the effect of d-electrons or the effect of dispersion forces. On the other hand flourine has a relatively low value of the dissociation energy owing to the repulsion between the nuclei which are not well shielded.

Rates of Reaction

The full discussion of the theories of the rates of chemical reactions is a major task which we shall not attempt. Our purpose in this section will be to show, in outline, how our earlier considerations about the way particles distribute themselves in energy levels can help to interpret the rates of chemical reactions.

The starting point for the discussion of reaction rates is the Arrhenius equation:

$$k_r = A \exp(-E_{\text{act}}/RT).$$

In this equation k_r is the velocity constant, E_{act} is the activation energy and A is a frequency factor. The activation energy is the energy barrier the reactants must pass over in order to become products. If we consider the general reaction

$$\text{X--Y} + \text{Z} \longrightarrow \text{X} + \text{Y--Z} \tag{5.2}$$

there will be an intermediate arrangement of the groups X,Y and Z in which the bond between X and Y is partially broken whilst the bond between Y and Z is only partially formed. We can represent this as $X \cdots Y \cdots Z$. This bonding arrangement is of higher energy than either the reactants $(XY + Z)$ or the products $(X + YZ)$. In the configuration of X, Y, and Z which has the highest energy the system is said to be an 'activated complex' or to be in its 'transition state'. The activation energy is then the extra energy of the transition state above the reactants. The way in which the energy changes during reaction is illustrated in Fig. 5.6.

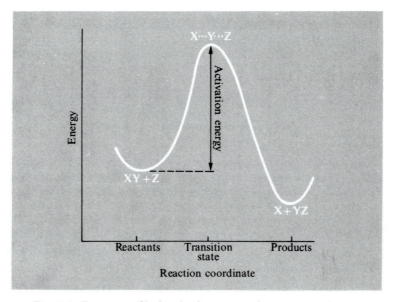

Fig. 5.6. Energy profile for the formation of a transition state.

The importance of thermodynamic considerations in reaction kinetics arises because it is usually assumed that the concentration of species in the transition state, $c^{\ddagger}$, is governed by an equilibrium between the reactant molecules and the transition state. We can then use the equilibrium constant $K^{\ddagger}$ for the reaction

$$XY + Z \longrightarrow X \cdots Y \cdots Z$$

Reactants   Transition
state

to calculate $c^{\ddagger}$:

$$c^{\ddagger} = K^{\ddagger} c_{XY} c_Z . \tag{5.3}$$

It should be noted that the postulate we have made is that $c^{\ddagger}$ is determined by Eq. (5.3), but not that there is necessarily a dynamic equilibrium in the ordinary sense between $c^{\ddagger}$ and the reactants. For simple reactions it is assumed that the transition state once formed is committed to becoming products. However, for complex reactions the probability that the transition state goes over to products may be less than unity. This complication is dealt with by introducing a 'transmission coefficient' for the reaction. The theory of reaction rates based on the equilibrium postulate is called 'transition-state theory' and sometimes, 'activated-complex theory'. We shall next show how this theory leads to an expression for k_r.

Transition State Theory

The overall rate at which the transition state $X \cdots Y \cdots Z$ decomposes (to $X + YZ$) is the product of its concentration $c^{\ddagger}$ and the frequency with which it decomposes $\nu^{\ddagger}$. This product is also the observed rate of the chemical reaction so we can write

$$\text{Rate} = \nu^{\ddagger} c^{\ddagger} . \tag{5.4}$$

In this equation $c^{\ddagger}$ is defined by Eq. (5.3) and the frequency $\nu^{\ddagger}$ remains to be determined. To find $\nu^{\ddagger}$ we must consider the physical and chemical nature of the transition state. Most of its properites are those of a conventional molecule, so that in principle we are justified in writing partition functions for translation, rotation, and vibration just as we would for a stable molecule. There is, however, one important way in which the transition state differs from a stable molecule, namely in the motion along the reaction coordinate. It is, of course, this motion that leads to the reaction. We need, therefore, to consider the partition function for this motion separately. We can look on the motion in two ways, each of which enables us to determine $\nu^{\ddagger}$. The starting point for both treatments is the equation we had earlier in this chapter (5.1) for the equilibrium constant K_N:

$$K_N = \frac{\text{Product of partition functions of products}}{\text{Product of partition functions of reactants}} e^{-\Delta E / RT} .$$

Thus for the equilibrium between reactants and transition state we write the equilibrium constant as:

$$K^{\ddagger} = \frac{q^{\ddagger}}{q_{XY} \cdot q_Z} \cdot \exp(-E_{\text{act}}/RT)$$

where $q^{\ddagger}$ is the total partition function for the transition state, or

$$K^{\ddagger} = \frac{q_{\ddagger} \cdot q_{\text{rc}}}{q_{XY} \, q_Z} \cdot \exp(-E_{\text{act}}/RT) \qquad (5.5)$$

where $q_{\ddagger}$ is a modified partition function which differs from $q^{\ddagger}$ in that the contribution of movement of Y along the reaction coordinate, q_{rc}, has been expressed separately. In order to determine the numerical value of $q_{\ddagger}$ we need to know the molecular parameters of the transition state. These are not available from experiment nor can they be calculated reliably. We can, therefore, only make order-of-magnitude estimates of the contributions to $q_{\ddagger}$, based on general chemical knowledge.

We shall now consider two ways in which an expression for q_{rc} can be obtained. These both depend on the postulate that once the transition state is formed there is no hindrance to it proceeding along the direction of the reaction coordinate.

Method 1. In this derivation we look on the motion in the direction of reaction as a vibration which has become very loose indeed so that the fundamental frequency has become extremely low. To see how this can happen we start by considering a modified potential-energy diagram in which there is a hollow at the top of the curve. An intermediate compound with quantized vibrational energy levels rather than the true transition state can then be formed. For this equilibrium between reactants and intermediates a true equilibrium constant can be written. We now consider what happens as this hollow is gradually filled in until eventually the top of the curve becomes convex. When this has happened only the transition state can exist at the top of the barrier. This sequence is illustrated schematically in Fig. 5.7(a, b, c), illustrating the change from the formation of an intermediate compound (a) through a very unstable intermediate (b) to a transition state (c).

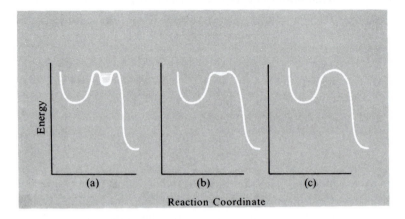

Fig. 5.7. The change from the formation of an intermediate compound (a) through a very unstable intermediate (b) to a transition state (c).

For the potential-energy curve illustrated in Fig. 5.7(a) we can write for the equilibrium between $(XY + Z)$ and the intermediate

$$K_N = \frac{q_i}{q_{XY}q_Z}\mathrm{e}^{-\Delta E/RT}$$

$$= \frac{q_i'(q_{\text{vib}})_i}{q_{XY}q_Z}\mathrm{e}^{-\Delta E/RT}$$

where q_i is the total partition for the intermediate and q_i' is this partition function without the contribution for vibration of the intermediate along the direction of the reaction coordinate $(q_{\text{vib}})_i$. For this vibration we can write the usual expression for a vibrational partition function

$$(q_{\text{vib}})_i = \frac{1}{1 - \mathrm{e}^{-h\nu/kT}}\;.$$

Now as we fill in the hollow and pass from Figs. 5.7(a) to 5.7(b), this vibration becomes progressively looser and the vibrational energy levels get closer together until, finally, when we have reached Fig. 5.7(c), a vibration once started along the reaction coordinate encounters no restoring force. Such a vibration then leads the completion of the reaction. For $(q_{\text{vib}})_i$ we are now justified in making the assumption that $h\nu \ll kT$. Remembering that the exponential in $(q_{\mathrm{vib}})_i$ can be expanded as

$$e^{-h\nu/kT} = 1 - \frac{h\nu}{kT} + \frac{1}{2}\left(\frac{h\nu}{kT}\right)^2 \cdots$$

we deduce that the partition function for vibration of the transition state along the reaction coordinate (which has frequency $\nu^{\ddagger}$) is

$$q_{\mathrm{rc}} \approx \frac{1}{1 - (1 - h\nu^{\ddagger}/kT)}$$

$$= \frac{kT}{h\nu^{\ddagger}}.$$

Substituting this equation in (5.5):

$$K^{\ddagger} = \frac{q_{\ddagger}(kT/h\nu^{\ddagger})\exp(-E_{\mathrm{act}}/RT)}{q_{XY}\,q_Z}.$$

Therefore

$$\frac{c^{\ddagger}}{c_{XY}c_Z} = \frac{q_{\ddagger}(kT/h\nu^{\ddagger})\exp(-E_{\mathrm{act}}/RT)}{q_{XY}q_Z}.$$

For this reaction we can write

$$\text{Rate} = k_r c_{XY} c_Z = \nu^{\ddagger}c^{\ddagger} = c_{XY}c_Z \cdot \frac{kT}{h} \cdot \frac{q^{\ddagger}}{q_{XY}q_Z}\exp(-E_{\mathrm{act}}/RT).$$

$$k_r = \frac{kT}{h}\frac{q^{\ddagger}}{q_{XY}q_Z}\exp(-E_{\mathrm{act}}/RT).$$

We have thus been able to use our expression for q_{rc} to derive an Arrhenius-like equation. We can also obtain some insight into factors which affect the magnitude of the pre-exponential factor A, since

$$A = \frac{kT}{h}\frac{q_{\ddagger}}{q_{XY}q_Z}. \tag{5.6}$$

We now turn to the second method of obtaining q_{rc}.

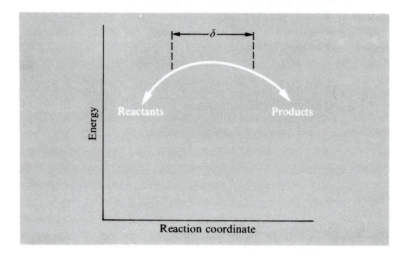

Fig. 5.8. Magnified view of a transition state.

Method 2.   This time we recognize from the beginning that there is no hindrance to the motion of Y along the reaction coordinate. The implication is then that we are dealing with a motion which is essentially a translation. Let us consider any small distance δ at the top of the barrier, as in Fig. 5.8. Then the (one-dimensional) translational partition function for the transition state along the length δ is

$$q_{\text{rc}} = \frac{(2\pi m^{\ddagger} kT)^{\frac{1}{2}} \delta}{h}$$

where $m^{\ddagger}$ is the effective mass of the transition state. Substituting as before:

$$K^{\ddagger} = \frac{c^{\ddagger}}{c_{XY} c_Z} = \frac{q_{\ddagger}}{q_{XY} \, q_Z} \frac{(2\pi m^{\ddagger} kT)^{\frac{1}{2}} \delta}{h} \cdot \exp(-E_{\text{act}}/RT) \,.$$

The kinetic theory can be used to calculate the average velocity with which the particles in the transition state move in the direction of reaction. The result is

$$v = \left(\frac{kT}{2\pi m^{\ddagger}} \right)^{\frac{1}{2}} \,.$$

The frequency with which particles cross the barrier is then v/δ. The rate of reaction is, as before:

$$\text{Rate} = \text{frequency of crossing} \times c^{\ddagger}$$

$$= \frac{vc^{\ddagger}}{\delta}$$

$$\text{Rate} = \left(\frac{kT}{2\pi m^{\ddagger}}\right)^{\frac{1}{2}} \frac{1}{\delta} \cdot c_{XY}c_{Z} \cdot \frac{q_{\ddagger}(2\pi m^{\ddagger}kT)^{\frac{1}{2}}\delta}{q_{XY} \cdot q_{Z}h} \exp(-E_{\text{act}}/RT)$$

but also

$$\text{Rate} = k_{r}c_{XY}c_{Z}$$

$$k_{r} = \frac{kT}{h} \frac{q_{\ddagger}}{q_{XY}q_{Z}} \exp(-E_{\text{act}}/RT).$$

The two derivations of k_{r} have thus, satisfactorily, led us to the same expression for k_{r}.

The factor kT/h has the units of time^{-1} and is thus a frequency. It is independent of the particular reaction studied and can be calculated at any chosen temperature from the tabulated values of k and h, e.g. at 300 K

$$kT/h = (1.38 \times 10^{-23})300/6.63 \times 10^{-34}s^{-1}$$

$$kT/h = 0.624 \times 10^{13}s^{-1}.$$

The partition-function ratio naturally varies greatly from one reaction to another depending on the molecular properites of the reactants and the transition state. We find, for example, that for the reaction between the simplest species, a pair of atoms, the value of A estimated from Eq. (5.6) is the same as the value calculated from the kinetic theory for the rate of collision of the atoms. Thus in the simplest case of the reaction between two structureless spheres both collision theory and transition-state theory yield the same expression. As the reactants become more complex, however, the calculated value of A becomes progressively lower than the rate of collision. This result is in general accord with experimental observations.

There is one further aspect of reaction rates which a consideration of the energy levels of the reactants and transition states can help to explain. This is the *kinetic isotope effect*. The basic observation is that for proton-transfer reactions the rate at which the proton reaction takes place is often faster than the rate of the deuteron reaction. A consideration of the energy levels of the reactants and the products shows why this should be so. Suppose that we are considering the transfer of proton from a carbon atom to a base. Schematically we can write the reaction as:

$$\diagdown\!\!\!\!-\text{C--H} + \,:\text{B} \rightarrow \diagdown\!\!\!\!-\text{C} \cdots \text{H} \cdots \text{B} \rightarrow \diagdown\!\!\!\!-\text{C} : ^{\ominus} + \text{H--B}^{\oplus}$$

acid   base   transition   base   acid
                           state

At normal temperatures the C–H bond of the reactant 'acid' will be in its ground vibrational energy level and thus have a zero-point energy $(1/2h\nu)_{\text{C--H}}$. Now suppose that the reaction takes place by the stretching of this bond and its eventual rupture with the simultaneous formation of a bond from the proton to the base B, i.e. the movement of the proton from acid to base is the reaction coordinate. In the transition state the proton is rather weakly bonded to both A and B, indeed if it starts to vibrate in the direction of B there is no restoring force and the reaction is completed. The effect of this is that we can neglect the zero-point energy of this vibration in the transition state. When we consider the deutero compound, the greater mass of the deuterium nucleus reduces the fundamental vibration frequency and, therefore, the zero-point energy $(1/2h\nu)_{\text{C--D}}$. It is now a bigger energy jump from the ground vibrational state to the transition state than for the proton-containing compound since the two compounds have a common transition state. On a simple energy-level diagram we can see this (Fig. 5.9).

We can make a rough estimate of the effect expected from the substitution of D for H using the Arrhenius equation.

$$k_H = A \exp(-E_H/RT)$$

$$k_D = A \exp(-E_D/RT)$$

$$\ln k_H/k_D = \frac{E_D - E_H}{RT}.$$

A typical C–H stretching vibration frequency is ~ 3000 cm^{-1}, for which the corresponding C–D frequency is roughly $3000 \times \sqrt{(\mu_{\text{C--H}}/\mu_{\text{C--D}})} = 2100$ cm^{-1}. Thus $E_H - E_D = \Delta$ (zero-point energies) $= \frac{1}{2}$ 900 cm^{-1} or approximately 5 kJ mol^{-1}. We can now calculate the relative rate of the two reactions at room temperature

$$\ln k_H/k_D = \frac{5000}{2500}$$

$$k_H/k_D \approx 7.5.$$

This is only an approximate calculation but it does illustrate the profound effect isotopic substitution may have. Naturally this effect is most noticeable for the isotopes of hydrogen because of the large ratio of the isotopic masses.

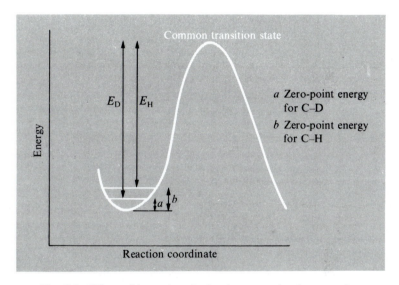

Fig. 5.9. Effect of isotopic substitution on activation energies.

Chapter 6

HEAT CAPACITIES

The importance of heat capacities in thermodynamic discussions can scarcely be over-estimated. At constant volume, the heat capacity measures quantitatively the ability of the system to take energy into its internal degrees of freedom. These are, in turn, intimately related to the atomic and molecular characteristics of the particular system. The heat capacity can thus provide an important link between the observed, macroscopic behaviour of a system and its detailed atomic or molecular structure.

Before we consider specific examples, let us look at the general way in which the energy-level ladder of an aspect of the system influences the ability of that aspect to take in energy, i.e. its heat capacity. When the levels are closely spaced and the energy gap between them is small, application of the distribution law in the form

$$n_i = n_0 p_i \exp(-\varepsilon_i / kT)$$

shows that it will be easy for particles to leave the ground state. In order to raise the temperature it is necessary to promote many particles to upper energy levels. There will be a large intake of energy and a heat capacity near the classical value which was deduced without considering quantization of energy. Conversely, when the energy levels are widely spaced and the energy gaps large, a rise in temperature promotes rather few particles and the heat capacity is low. Now the description 'large' or 'small' for the spacings of the energy levels are of necessity comparative; any quantity or object is only large or small compared with something else. In this discussion, the comparison is with thermal energy, kT. We can illustrate this with two important physical situations:

$\varepsilon \ll kT$ (when the heat capacity has its classical value)

and

$\varepsilon \gg kT$ (when the heat capacity tends to zero)

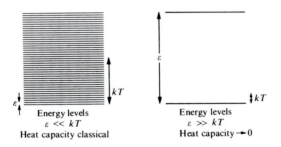

Fig. 6.1. Effect of energy-level spacing on heat capacities.

as shown in Fig. 6.1. Put another way, when the energy separation is much less than kT it is as though the effect of quantization were not noticed by the thermal energy. The classical result, which is based on a continuous distribution of energy levels, is then obtained. But once the energy becomes of the order of or greater than kT, quantization effects become important. We shall now consider two important systems, the study of which can help greatly in understanding the effect of quantization of energy on the thermodynamic properties of substance.

Heat Capacities of Gases

When a simple gas is heated at constant volume the energy is taken up by the modes of the molecule: translation, rotation, and vibration. Although translation energy is quantized, the quanta are so small ($\varepsilon_{trans} \ll kT$) that under all experimentally accessible conditions the translational partition function can be evaluated explicitly by integration. As we saw in Chapter 4, the energy of translation motion calculated from the partition function, is:

$$U_{trans} = \frac{3}{2}RT.$$

Therefore

$$C_{\text{trans}} = \frac{dU_{\text{trans}}}{dT} = \frac{3}{2}R.$$

No deviations from this formula due to quantization effects have been detected.

The energy levels of a linear rotating molecule are given by the formula discussed in Chapter 4

$$E_{\text{rot}} = \frac{h^2}{8\pi^2 I}J(J+1)$$

where J is a positive integer or zero. From this we deduced in Chapter 4 that, for diatomic molecules other than the isotopes of hydrogen, to a very good approximation:

$$q_{\text{rot}} = \frac{8\pi^2 IkT}{\sigma h^2}.$$

For a linear molecule:

$$U_{\text{rot}} = RT^2 \frac{d \ln q_{\text{rot}}}{dT} = RT.$$

Therefore,

$$C_{\text{rot}} = \frac{dU_{\text{rot}}}{dT} = R.$$

(This result is identical with the classical formula in which the two degrees of rotational freedom of a linear molecule each contribute $\frac{1}{2}R$ to C_{rot}.) In practice we find that for all diatomic gases, except hydrogen and its isotopes, the experimental value is indeed R and quantization does not affect the rotational heat capacity. That quantization of rotation has the greatest effect for hydrogen should not be surprising. The spacing of rotational energy levels is inversely proportional to the moment of inertia, with the result that the energy separation of $J = 0$ and $J = 1$ is greatest for hydrogen. For example, this spacing is 30 times as large in hydrogen as it is in nitrogen. Thus, whilst quantization effects are noticeable for hydrogen between 20 K and 300 K, the temperature would need to be lowered by a factor of 30 to produce similar effects in nitrogen. The highest temperature at which non-classical behaviour could be expected from nitrogen is thus 10 K, at which temperature it is frozen.

The vibrational energy levels of a diatomic simple harmonic oscillator are given the formula:

$$E_{\text{vib}} = h\nu(v + \frac{1}{2})$$

where ν is the frequency of the vibration and v is the vibrational quantum number, a position integer or zero. As we saw in Chapter 4 this leads to the vibrational partition function

$$q_{\text{vib}} = (1 - e^{-h\nu/kT})^{-1}$$

$$\ln q_{\text{vib}} = -\ln(1 - e^{-h\nu/kT})$$

$$\frac{d(\ln q_{\text{vib}})}{dT} = \frac{h\nu}{kT^2} \cdot \frac{e^{-h\nu/kT}}{1 - e^{-h\nu/kT}}$$

$$U_{\text{vib}} = RT^2 \frac{d(\ln q_{\text{vib}})}{dT} = RT \cdot \frac{h\nu}{kT} \cdot \frac{e^{-h\nu/kT}}{1 - e^{-h\nu/kT}}.$$

When the substitution $u = h\nu/kT$ is made, this equation is of the form quoted in Appendix 3. If this equation is rearranged we can obtain:

$$U_{\text{vib}} = \frac{Rh\nu}{k} \cdot \frac{1}{e^{(h\nu/kT)} - 1}$$

$$C_{\text{vib}} = R\left(\frac{h\nu}{kT}\right)^2 \frac{e^{h\nu/kT}}{(e^{h\nu/kT} - 1)^2}. \qquad (6.1)$$

With the substitution $u = h\nu/kT$, this equation is also quoted in Appendix 3.

The vibrational heat capacity of a diatomic molecule, calculated from Eq. (6.1), can vary between zero and R, depending upon the value of the ratio $h\nu/kT$ (see Appendix 3). At high temperatures when $h\nu/kT$ is small, we can expand $(1 - e^{-h\nu/kT})^{-1}$ as a power series and neglect terms in $(h\nu/kT)^2$ and beyond. We then get:

$$q_{\text{vib}} = \frac{kT}{h\nu} \text{(high temperature)}$$

$$C_{\text{vib}} = \frac{d}{dT}\left(RT^2 \frac{d \ln q_{\text{vib}}}{dT}\right),$$

whence

$$C_{\text{vib}} = R.$$

At low temperatures where $h\nu/kT$ is large, $e^{-h\nu/kT} \to 0$ and $q_{\text{vib}} \to 1$. Hence

$$C_{\text{vib}} \to 0 \text{ as } T \to 0.$$

The frequency ν is related to the mass and bond strength of the molecule by the equation $\nu = \frac{1}{2\pi}\sqrt{(k/\mu)}$, where k is the force-constant of the bond and μ the reduced mass of the molecule. Thus light (low μ) strongly bonded (high k) molecules have high vibration frequencies and low vibrational heat capacities at room temperature. Conversely, heavy, weakly-bonded molecules have near-classical vibration heat capacities, e.g. at 300 K C_{vib} ~ 0 J K^{-1}mol^{-1} for H_2; ~ 4.2 J K^{-1}mol^{-1} for Cl_2, and ~ 8.4 J K^{-1}mol^{-1} for I_2.

It is interesting to note that a ten-fold change in the ratio $h\nu/kT$ from a low-temperature value of, say 6 to a high-temperature value of 0.6 has the effect of changing the heat capacity from $0.09R$ to $0.97R$.

The temperature dependence of the heat capacity of a diatomic gas can be illustrated schematically (Fig. 6.2).

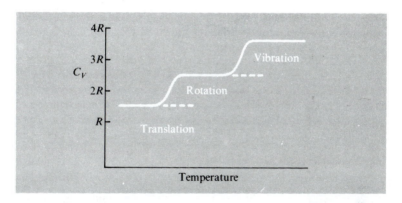

Fig. 6.2. Variation of heat capacity of a diatomic gas with temperature (only hydrogen isotopes give purely translation heat capacities).

Heat Capacity of Simple Solids

One consequence of the third law is a prediction that the heat capacity of a solid tends to zero at very low temperatures. We can see how this happens if we note that the only way in which a monatomic solid can take in heat is by increasing the vibrational excitation of its constituent atoms. This excitation naturally increases the entropy of the system. When the temperature is lowered sufficiently all the particles fall back to the lowest available level, thus reducing the entropy to zero. As we have seen for the diatomic molecule, when there is no vibrational excitation ($h\nu \ll kT$)

the vibrational heat capacity tends to zero. This result can be expressed mathematically for the heat capacity at constant pressure as follows:

$$C_P = \left(\frac{\partial H}{\partial T}\right)_P \text{ by definition}$$

$$= \left(\frac{\partial q}{\partial T}\right)_P \text{ (since only } P - V \text{ work is done by the system)}$$

$$= \left(T\frac{\partial S}{\partial T}\right)_P \text{ (a reversible process)}$$

$$= \left(\frac{\partial S}{\partial \ln T}\right)_P.$$

Now as T tends to zero, the third law predicts that the entropy tends to zero. Since $\ln T$ tends to minus infinity as $T \to 0$, the bracket tends to zero, i.e. $C_P \to 0$. Similarly $C_V \to 0$ as $\to T \to 0$. The temperature dependence of C_V of for some monatomic solids is shown in Fig. 6.3.

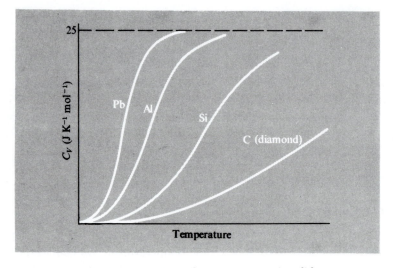

Fig. 6.3. Heat capacities of some monatomic solids.

The high temperature limiting value of C_V for a monatomic solid is $3R$, about 25 J K^{-1}mol^{-1}. This is, of course, Dulong and Petit's Law. Classically it was derived by assuming each atom could vibrate in three directions, with each mode of vibration contributing R to the heat capacity.

This approach, however, could provide no explanation for a heat capacity which goes to zero.

In order to understand the behaviour of solids, we need to find a model for the system which seems physically reasonable and whose thermodynamic properties are in accord with the experimental results. There have been two particularly important approaches to this problem, first by Einstein and then by Debye.

The Einstein Theory of Heat Capacities

In this treatment, the model is of a solid which consists of N independent simple harmonic oscillators whose energy is quantized according to the equation $E = nh\nu$, where ν is the fundamental frequency of the oscillators and n is a positive integer or zero, i.e. an oscillator may only have discrete energies $E, 2E, 3E \ldots$ above the zero-point level. It is at this point that the model differs from the classical treatment, in which the oscillating atoms can have any frequency and, therefore, any energy. When there is no constraint on the vibrations the solid has a temperature-independent heat capacity of $3R$.

The problem we now face is one of finding the energy of a system of solid-state simple harmonic oscillators. It can be solved by the same mathematical procedure as that we have already used earlier in this Chapter for the vibrations of a diatomic molecule. The only point of difference arises because a gaseous diatomic molecule has but a single mode of vibration, along the internuclear axis, while the motion of a solid-state oscillator can be resolved into three components, one along each of the Cartesian coordinates. Thus from our earlier equation for a vibrating diatomic molecule:

$$U_{\text{vib}} = \frac{Rh\nu}{k} \cdot \frac{1}{e^{h\nu/kT} - 1}$$

we can calculate the average energy of an oscillator $\bar{E}$ using the relation

$$\bar{E} = \frac{U}{N}.$$

Hence

$$\bar{E} = \frac{h\nu}{e^{h\nu/kT} - 1} \qquad (R = kN).$$

Since each solid-state oscillator has 3 directions of vibration or 'degrees of freedom' the total vibrational energy $3N\bar{E}$. We can now reach the Einstein value of the heat capacity of differentiation:

$$U_E = 3N\bar{E} = 3N\frac{h\nu}{e^{h\nu/kT} - 1}$$

$$C_E = \left(\frac{\partial U}{\partial T}\right)_V = \frac{3Nk\left(\frac{h\nu}{kT}\right)^2 e^{h\nu/kT}}{\left(e^{h\nu/kT} - 1\right)^2}.$$

At high temperature ($h\nu \ll kT$) the upper limit of C_E can be obtained by the same method as the one used for a gas. The result is

$$(C_E)_{\text{High } T} = 3R.$$

At low temperature ($h\nu \gg /kT$) the energy is:

$$U_E = 3Nh\nu e^{-h\nu/kT}.$$

Therefore

$$(C_E)_{\text{Low}T} = 3Nk\left(\frac{h\nu}{kT}\right)^2 e^{-h\nu/kT}.$$

As $T \to 0$ the exponential term controls the equation and $C_V \to 0$. This again is the correct limit. However, the exponential decrease at the lowest temperature is more rapid than the experimental results, which generally have a limiting T^3 dependence.

In the equation for C_V the only unknown parameter is the frequency of the oscillators, ν. Thus, by fitting the equation as well as possible to the experimental results, a value of this characteristic frequency of any particular solid can be obtained. It is sometimes more convenient to use temperature than frequency as the characteristic parameter. The two are related by the equation

$$h\nu_E = k\theta_E$$

where θ_E is the Einstein temperature.

For many common inorganic crystals $\theta_E \sim 200K$, so that $\nu_E \sim 4 \times 10^{12}$Hz. It is usually possible to get good agreement between theory and experiment down to a temperature of about $0.2\theta_E$. Below this temperature the exponential factor causes too rapid a fall in the heat capacity.

It is useful to consider the shortcoming of the Einstein model. The postulate of a single vibration frequency for the particles appears to be a good approximation at intermediate and high temperatures ($T > 0.2\theta_{\rm E}$). But at low temperatures the spacing of the lowest energy levels, which is $h\nu_{\rm E}$, is too large. The result is that the heat capacity falls too rapidly. This conclusion illustrates our general observation that widely spaced energy levels, or large quanta, lead to a small heat capacity. In choosing a new model, therefore, we should seek one in which there are some, but not too many, low-lying energy levels, which can be populated at low temperatures. The improvement at low temperatures which the Debye treatment offers over the Einstein model arises from this feature.

The Debye theory of heat capacities

Instead of treating the particles as though the motions were independent of one another, the Debye approach recognizes that the particles do interact. In this model the solid is treated as a homogeneous continuum, and the allowed vibrational energy levels become those of the crystal as a whole. The vibrational motion can be thought of as the three dimensional analogue of the familiar vibrations of a violin string. The quantization of energy is then governed by the condition that the vibrations have a node at the edge of the solid, i.e. $l = n\lambda/2$ where l is the length of the side of the solid and λ the wavelength of the vibration. n is a positive integer.

In order to calculate the total energy of vibration we need to know how many oscillations there are at each allowed frequency. We will call the mathematical function which gives us this result $f(\nu)$. The elastic theory of solids can be used to calculate $f(\nu)$, with the result $f(\nu) \propto \nu^2$. Over any small range of frequency, at any particular frequency ν_i, the number of vibrations is $f(\nu_i)\delta\nu$. Since we still have N atoms, the overall number of allowed vibrations remains at $3N$, i.e.

$$\sum_{\nu} f(\nu)\delta\nu = 3N \,.$$

An important consequence of this equation is that $f(\nu)$ and, therefore, ν cannot go on increasing indefinitely but must reach a limiting value. This limiting frequency is called the *Debye cut-off frequency*, $\nu_{\rm D}$. In this model the energy levels of the vibrating solid are closely spaced, so that no serious mathematical error is introduced if we integrate instead of summating the equation for $f(\nu)$, i.e.

$$\int_0^{\nu_D} a\nu^2 d\nu = 3N$$

where a is a constant of proportionality.

Hence we get

$$a = 9N/\nu_D^3, \quad f(\nu) = 9N\nu^2/\nu_D^3.$$

We are now ready to write down the total vibrational energy of the crystal. This energy is the product of the average energy of an oscillator of frequency ν and the number of oscillators at that frequency, summed or integrated over all the allowed frequencies. This latter range is effectively 0 to ν_D. The average energy of an oscillator we have calculated previously in the Einstein treatment; it is

$$\bar{E} = \frac{h\nu}{e^{h\nu/kT} - 1}.$$

Thus

$$U = \int_0^{\nu_D} \bar{E} f(\nu) d\nu$$

$$= \int_0^{\nu_D} \frac{h\nu}{e^{h\nu/kT} - 1} \frac{9N\nu^2 d\nu}{\nu_D^3}.$$

Once again we can express the characteristic frequency ν_D as a temperature, θ_D, by means of the equation $h\nu_D = k\theta_D$. If we let $u = h\nu/kT$ and differentiate U to get C_V the result is:

$$C_D = 9R\left(\frac{T}{\theta_D}\right)^3 \int_0^{\theta_D/T} \frac{u^4 e^u du}{(e^u - 1)^2}$$

This integral can be calculated and is tabulated in standard works; at low temperatures the limit θ_D/T on the integral may be replaced by ∞, the value of the integral becomes independent of T, and so the heat capacity varies as T^3/θ_D^3. Once again C_V depends on the ratio of the temperature of a characteristic temperature, this time θ_D. This latter quantity is the only parameter specific to any particular monatomic solid. It follows that if θ_D is chosen correctly and C_V is plotted against T/θ_D all solids should fall on the same curve. This should also be true of the Einstein model if C_V is plotted against T/θ_E. The experimental results show that this expectation is rather accurately fulfilled but that, as we have seen, the Einstein heat

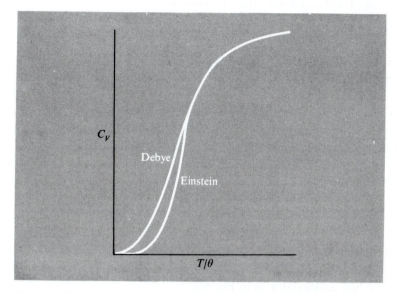

Fig. 6.4. Comparison of Einstein and Debye theories of heat capacities.

capacity falls off too fast at low temperatures. The Debye and Einstein curves are shown in Fig. 6.4.

Having obtained θ_D from the curve it is possible to make a further test of the Debye theory, because the elastic theory of solids allows a quite independent calculation of θ_D to be made, using only the independently measured elastic properties of the solid. The relation is:

$$\theta_D = \frac{h}{k} \cdot V_0 \left(6\pi^2 \frac{N}{V} \right)^{\frac{1}{3}}$$

where V_0 is the velocity of sound in the solid of volume V. There is good, though not perfect, agreement between the two results. The discrepancies arise because the distribution of allowed vibration frequencies chosen in the Debye treatment is something of a simplification. A more elaborate treatment gives better agreement but a less useful formula.

A particularly valuable feature of the Debye equation is the simple form of the low-temperature region. This T^3 temperature dependence can be used to extrapolate the experimental results from the lowest accessible temperature to the absolute zero. As we shall see, this extrapolation is required for the experimental determination of the third-law entropies.

Effects of Electron on Heat Capacities

One distinctive property of metals is their electrical conductivity. This property results from the ability of the conduction electrons to move freely through the lattice. It might have been expected, therefore, that these electrons would make a significant contribution to the heat capacity of the metal. However, at room temperature the difference between the heat capacity of a metal and of a dielectric is hardly significant. By contrast, at very low temperatures, where the lattice heat capacity is low and falling rapidly (Debye T^3 region), the contribution of the electrons does become important and it can be measured. The equation for the total heat capacity in the Debye T^3 region is

$$C_{\mathrm{metal}} = \underset{\text{(lattice)}}{\alpha T^3} + \underset{\text{(electrons)}}{\gamma T} \quad.$$

Thus at low temperatures the electrons make an appreciable contribution to the heat capacity. This is most clearly seen plotting C/T against T^2 for a metal and an insulator as in Fig. 6.5.

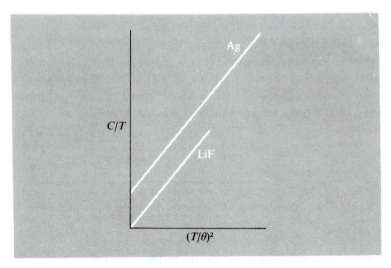

Fig. 6.5. Comparison of low-temperature heat capacities of a metal and an insulator.

The reason why the electrons have a heat capacity which is so much less that the classical value can be understood from our earlier discussions of the effect of the quantization of energy levels on heat capacities. We have seen that when the characteristic energy is large compared with kT the associated heat capacity is small. The elementary band theory of metals can be used to show how it comes about that the energy of the electrons is indeed much greater than kT.

If we imagine that the metal consists of a regular crystalline array of atoms with the electrons free to move throughout the lattice then we find that there are groups of contiguous energy levels, 'bands', and that these may be separated from neighbouring bands by an energy gap. Within each band, the electrons (being Fermi-Dirac particles) occupy the lowest energy levels consistent with the Pauli principle, i.e. only two electrons, with opposed spins, can occupy each energy level. Thus, as electrons are fed in to the lattice, higher and higher energy states must be occupied. The energy-level diagram for this process usually drawn by plotting the number of states with a particular energy, $N(E)$, against energy. The resulting curve is the band. Electrons then occupy the $N_e/2$ lowest energy levels, where N_e is the number of conduction electrons. The resulting situation at absolute zero is shown in Fig. 6.6. The area shown shaded represents energy levels occupied by electrons, the clear area represents empty levels and the vertical line E_F marks the boundary. E_F is called the *Fermi level.*

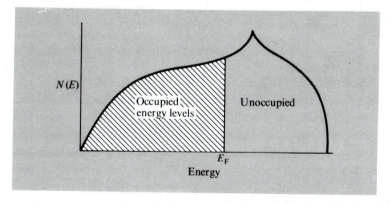

Fig. 6.6. A simple band diagram for a metal.

As the diagram shows, it is the electrons at the Fermi level and these alone which have unoccupied energy levels adjacent to them. Excitation to these empty levels may be achieved thermally or by the application of an electric field; in the latter case a current flows. However, the energy scale is a very coarse one and the energy at the Fermi level is typically ~ 5 eV or 520 kJ mol^{-1}. On such a scale thermal energy, kT, is always very small indeed at all ordinary temperatures. But only those electrons which lie within an energy range $\sim kT$ of the Fermi level can reach the empty levels and because $E_F \gg kT$, these electrons only form a very small fraction of the total number. Thus only a few electrons are capable of being excited and the electronic heat capacity is very low. In order to allow the electrons to make a nearly classical contribution to C_V we should require kT to become comparable with E_F, i.e. $T \sim 60\ 000$ K. Thus for electrons in metals, all temperatures are 'low' and the quantum mechanical result

$$C_{\text{el}} \propto T \text{ at low temperatures}$$

is always valid.

The Use of Heat Capacity Data in the Calculation of Entropy Changes

The third law of thermodynamics requires that for a system in equilibrium the entropy goes to zero at absolute zero. If then we measure the heat capacity and latent heats of any phase transitions from the absolute zero to any chosen temperature for each of the compounds involved in a chemical change

$$A + B \rightleftharpoons C + D$$

we can calculate $\Delta S (= S_C + S_D - S_A - S_B)$ for the reaction.

There may, however, be alternative classical methods of determining ΔS, some of which are described below. The result is the same whichever way ΔS is calculated and this, in part, justifies the assumption that $S_0 = 0$. Such comparisons are indeed often referred to as experimental justifications of the third law.

Examples of Determinations of Entropy Changes in Chemical Reactions

We can obtain standard entropy changes using the relationship

$$\Delta S^{\ominus} = (\Delta H^{\ominus} - \Delta G^{\ominus})/T .$$

$\Delta H^{\ominus}$ may be measured calorimetrically and $\Delta G^{\ominus}$ by means of e.m.f. measurements. The values of $\Delta S^{\ominus}$ may then be compared with that found from heat-capacity experiments. Good arrangement is normally obtained.

Typical examples of such comparisons which have been made are

$$Hg_{(l)} + AgCl_{(s)} = \frac{1}{2}Hg_2Cl_{2(s)} + Ag_{(s)}$$

and

$$\frac{1}{2}Pb_{(s)} + AgI_{(s)} = \frac{1}{2}PbI_{2(s)} + Ag_{(s)}$$

using cells such as

$$Hg; Hg_2Cl_2 \,|\, \text{chloride solution} \,|\, AgCl; Ag$$

for the first reaction. In this case $\Delta S^{\ominus}$ at 298 K was found to be 32.52 J K^{-1} mol^{-1} from the combination of $\Delta G^{\ominus}$ and $\Delta H^{\ominus}$ and 32.19 J K^{-1}mol^{-1} by use of the third law; a very satisfactory result.

The polymorphic change between rhombic (α) and monoclinic (β) sulphur has also been studied in a similar way. At the transition temperature the two forms are in equilibrium and $\Delta G^{\ominus} = 0$. The enthalpy change at this temperature can be determined by a direct calorimetric experiment, with the result at the transition temperature (368.6 K) $\Delta H^{\ominus} = 397 \pm 40$ J K^{-1}mol^{-1} Thus at 368.6 K

$$\Delta S^{\ominus} = -\Delta H^{\ominus}/T = 1.08 \pm 0.1 \text{ J K}^{-1}\text{mol}^{-1}.$$

By comparison the third-law value from heat capacities is

$$\Delta S^{\ominus} = 0.91 \pm 0.2 \text{ J K}^{-1}\text{mol}^{-1}.$$

Again the agreement is excellent.

There have been many experiments in which it has been possible to compare the entropy change measured directly with the value obtained from heat-capacity measurements made to low temperatures (extrapolated using the Debye T^3 relationship); for example, the entropy change of the solid-state phase transition between two forms of phosphine, and the entropies of many ions in aqueous solution. The latter values are relative to the hydrogen ion.

Comparison of Experimental and Calculated Individual Molecular Entropies

For reactions in gases and indeed for individual gases we can get striking verification of the third law from the agreement between entropies calculated from heat-capacity data and entropies calculated by means of partition functions. The information about the energy levels of the molecule needed for the calculation of partition functions is available with high accuracy from spectroscopic measurements. The heat-capacity data for N_2 show how the entropy of this gas at its boiling point is made up:

Temperature	*Entropy* $(J\ K^{-1}mol^{-1})$
$0 - 10$ K(Debye T^3)	1.91
$10 - 35.61$ K (solid)	25.22
phase transition(L/T)	6.42
$35.61 - 63.14$ (solid)	23.36
melting(L/T)	11.41
$63.14 - 77.32$ (liquid)	11.40
boiling(L/T)	72.05
Total	151.77 J K^{-1} mol^{-1}.

When the entropy of N_2 at its boiling point is calculated using the partition functions the value is 152.7 J $K^{-1}mol^{-1}$ if it is assumed that the gas is ideal. Exact agreement is obtained when a correction for non-ideality is made. Similar highly satisfactory results are usually obtained when calculated and measured entropies of simple gases are compared. However, when more complicated molecules are considered a fundamental difficulty may arise in the choice of the partition function to describe the intramolecular motion of groups of atoms relative to the rest of the molecules. Such a motion is usually called an internal rotation and is discussed in the next section.

Internal Rotation

In many molecules there is an identifiable group of atoms joined to the rest of the molecule by a single chemical bond; e.g. the methyl group in methylbenzene. The rotation of the group about this bond is called an internal rotation. If there is no restriction on the movement we say that it is free, but if there is energy barrier between different rotational positions of the group the motion is said to be *hindered*.

A rotation of the group through 360° returns the atoms to their original positions. The energy difference between the peaks and troughs of energy encountered by the group during this rotation forms the activation energy for the internal rotation. When the barrier becomes high enough to inhibit rotation almost entirely an angular displacement of the group when initially in a potential-energy minimum is followed by its return to the same position. This results in the group executing a rocking motion, which is neither a rotation nor strictly speaking a simple harmonic vibration. The movement is usually called a 'torsional oscillation' or 'libration' and it will have its own energy levels and statistical thermodynamic functions.

The way in which the contribution of internal rotation to the heat capacity of a gas varies with temperature can be calculated for any particular barrier height. When the barrier is a high one and the motion is a torsional oscillation the minimum heat capacity is zero and the maximum heat capacity is R. This upper value is reached only at high temperatures, e.g. for a barrier height of 42 kJ mol^{-1} a temperature above 600 K is needed. As the barrier is lowered the motion turns into hindered rotation and then to free rotation when the barrier is zero. In this latter case the heat capacity has the value characteristic of one degree of rotational freedom i.e. $R/2$. There is thus a significant difference between the upper limits, of R and $R/2$, in the extreme cases of a very high and a very low barrier. The temperature dependence of the internal rotation heat capacities in some intermediate cases is shown in Fig. 6.7. By comparison between theory and experiment it may be possible to deduce the height of the rotational barrier for any particular molecule.

An alternative, related approach is to calculate the entropy of the molecule assuming free rotation about the bond and to compare the result with experiment. When there is a discrepancy and the molecule is one in which it is plausible to attribute the discrepancy to hindered rotation, the entropy difference can be related to the barrier height. For example, the experimental entropy of gaseous perfluoropropane at its boiling point is 373 J K^{-1}mol^{-1} whereas the calculated entropy is 389 J K^{-1}mol^{-1}, made up of $S_{\text{trans}} = 170$ J K^{-1}mol^{-1}, $S_{\text{rot}} = 116$ J K^{-1}mol^{-1}, $S_{\text{vib}} = 49$ J K^{-1}mol^{-1} and $S_{\text{free rotn}} = 54$ J K^{-1}mol^{-1}. The discrepancy is thus only 16 J K^{-1}mol^{-1}. On recalculating the contribution from internal rotation but assuming a barrier of 12.6 kJ mol^{-1} exact agreement with experiment is obtained.

The magnitudes of the energy barriers are of great practical interest since it is such considerations that enable predictions of the shapes of molecules to be made. Conformational analysis has been an important topic in organic chemistry for a considerable time and this notion of deciding upon the shape of a molecule which results form non-bonded interactions is now of even greater interest in biological work. Many molecules, both small drug molecules and huge enzymes, owe their biological activity to their precise shape and their ability to adopt a particular conformation. As a result there have been many attempts to measure and understand barriers to internal rotation in these, as well as in smaller, molecules.

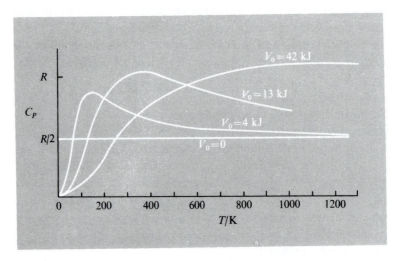

Fig. 6.7. Effect of barrier heights (V_0) on the temperature-dependence of the heat capacity due to internal rotation of an ethane-like molecule.

Chapter 7

LOW-TEMPERATURE PHENOMENA

High and Low Temperatures

We are accustomed to think of temperature either in terms of the expansion of a gas using the ideal-gas formula $PV = nRT$, so that at constant pressure $V_1/V_2 = T_1/T_2$, or in terms of the efficiency of a reversible thermodynamic machine. If such a machine operating between temperature T_1 and T_2 takes in heat Q_1 at T_1 and gives out heat Q_2 at T_2, then T_1 and T_2 are defined by

$$\frac{Q_2}{Q_1} = \frac{T_2}{T_1}.$$

These two concepts of temperature introduce the idea that it is not so much temperature *differences* that are of interest as temperature *ratios*. Thus the step from 1 K to 10^{-3} K is as significant as the step from 1000 K to 1 K.

Our statistical ideas support and extend this point of view when we consider the role of temperature in the Boltzmann distribution. If we consider a non-degenerate two-level system separated by an energy gap ε then

$$\frac{n_{\text{upper}}}{n_{\text{lower}}} = \text{e}^{-\varepsilon/kT}.$$

Let $\varepsilon = kT_c$. Then when $T = 5 \times T_c$

$$\frac{n_{\text{upper}}}{n_{\text{lower}}} = \text{e}^{-1/5} = 0.82.$$

The partition function $q = \sum e^{-\varepsilon/kT} = 1 + 0.82 = 1.82$. For this two-level system $q_{max} = 2$ and $q_{min} = 1$.

If, however, $T = 0.5T_c$

$$\frac{n_{upper}}{n_{lower}} = e^{-1/0.5} = 0.14 \text{ and } q = 1.14.$$

Thus at the higher temperature the particles are fairly equally distributed and the entropy is high, while at the lower temperature the particles are concentrated in the lower level and the entropy is low. We can calculate the maximum entropy of the system by considering the high temperature limit, where $\varepsilon \ll kT_c$. Under these conditions $n_{lower} \approx n_{lower}$ and any one particle has approximately equal chances of being in the upper and lower levels, i.e. the number of distributions per particle is 2. For N particles, therefore,

$$W = 2^N$$

and

$$S = k \ln W = R \ln 2.$$

Quite generally, when we refer to a high temperature we imply that some aspect of the system with which we are concerned has a considerable entropy. That is to say, in the Boltzmann expression from non-degenerate energy levels:

$$n = n_0 e^{-\varepsilon/kT},$$

the energy ε is considerably less than kT.

We should add that if it could be arranged for such a system, that $n_{upper} > n_{lower}$, then according to this expression we would have a 'negative temperature' or an infinite one if $n_{upper} = n_{lower}$. This is a point of some importance to which we will return the discussion of lasers (in Chapter 8), where it is of great practical significance.

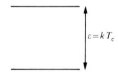

Fig. 7.1. A two-level energy system.

The heat capacity of the two-level system of Fig. 7.1 varies in an interesing way as the temperature is raised. At very low temperatures, where $\varepsilon \gg kT$, practically all the molecules are in the bottom level. For a modest rise in temperature, as we have already seen in Chapter 6, only a small proportion of the particles are excited to the first level and the heat capacity is low. At high temperatures where $\varepsilon \ll kT$ the populations of the two levels have become nearly equal. When this has happened no further significant promotion of particles is possible, the system can take in no more energy, and the heat capacity again tends to zero. At the high temperature limit we can look on the system as having become 'saturated' with energy. The effect of 'saturation' can sometimes be observed in spectroscopy if the exciting radiation is applied with such high intensity that particles get promoted from the lower level to the upper more rapidly than they can drop back from the upper to the lower. This frequently occurs in nuclear magnetic resonance spectroscopy of protons in solids. We have now established that both the low temperature and high temperature limits of the heat capacity of a two-level system are zero, from which it necessarily follows that there must be an intermediate temperature at which heat capacity passes through a maximum.

We should note here that the essential difference between the behaviour of the two-level system and the systems we have discussed previously, is that for these latter systems we have always assumed that an unlimited number of energy levels is available. However hot we make the system there are then always yet higher energy levels available into which particles can be promoted. The heat capacity then retains its non-zero, classical high-temperature value. By contrast, the capacity of any system with a limited number of energy levels for taking energy will be limited. Its heat capacity versus temperature curve will have the same general form as we have described for a two-level system. This form of temperature dependence of heat capacity is often known at the Schottky effect.

We can explore the Schottky effect further if we consider a more general situation in which the degeneracies of the two levels are no longer unity but are p_0 and p_1 for the lower and upper levels respectively. Now as we saw in the Einstein theory of heat capacities (Chapter 6) the average energy of a particle is given by

$$\bar{E} = U / \sum_i n_i$$

and

$$U = \sum_i n_i \varepsilon_i .$$

Then if we substitute in the Boltzmann equation in the form

$$n_i = n_0 \, p_i \exp(-\varepsilon_i/kT)$$

we obtain, for the two-level system of Fig. 7.1

$$\bar{E}_{2\text{-level}} = \frac{\varepsilon p_1 \exp(-\varepsilon/kT)}{p_0 + p_1 \exp(-\varepsilon/kT)} .$$

For one mole, therefore, the total energy is

$$U_{2\text{-level}} = \frac{N\varepsilon \, p_1 \exp(-\varepsilon/kT)}{p_0 + p_1 \exp(-\varepsilon/kT)} = \frac{N\varepsilon \, p_1 \exp(-T_c/T)}{p_0 + p_1 \exp(-T_c/T)} . \tag{7.1}$$

The heat capacity of a two-level system is then

$$(C)_{2\text{-level}} = (\partial U_{2\text{-level}}/\partial T)$$

and differentiation of (7.1) yields:

$$(C)_{2\text{-level}} = R\left(\frac{T_c}{T}\right)^2 \cdot \frac{(p_0/p_1)\exp(T_c/T)}{\{1 + (p_0/p_1)\exp(T_c/T)\}^2} . \tag{7.2}$$

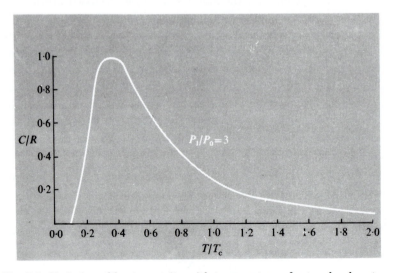

Fig. 7.2. Variation of heat capacity with temperature of a two-level system.

This equation is plotted in Fig. 7.2 for $p_1/p_0 = 3$. We can see that Eq. (7.2) is of the correct form because at high temperatures the exponential terms tend to unity and the $(T_c/T)^2$ factor becomes very small. At low temperatures the ratio of the exponential terms tends to $p_1/p_0 \exp(-T_c/T)$ which declines more rapidly than $(T_c/T)^2$ increases and again $(C_v)_{\text{2-level}} \to 0$.

The most important effect of increasing the degeneracy of the upper state relative to the lower is that the value of $(C_v)_{\text{2-level}}$ at its maximum increases markedly. We can deduce that this is so by differentiating Eq. (7.2) and remembering that, at the maximum,

$$\frac{dC_{\text{2-level}}}{dT} = 0 \,.$$

The differentiation is straightforward and the equations for the temperature of the maximum, $T_{\max}$ and for the maximum value of $C_{\text{2-level}}$ are:

$$\frac{T_c/T_{\max} + 2}{T_c/T_{\max} - 2} = \frac{p_0}{p_1} \exp(T_c/T_{\max})$$

and

$$(C_{\text{2-level}})_{\max} = R\{(T_c/T_{\max})^2 - 4\}/4 \,.$$

The results, for a range of values of p_1/p_0 are listed in Table 7.1.

Table 7.1.

p_1/p_0	$T_c/T_{\max}$	$(C_{\text{2-level}})_{\max}$
1	2.40	0.44 R
2	2.65	0.76 R
3	2.84	1.02 R
4	3.00	1.25 R
5	3.12	1.44 R

These considerations of a two-level system can help us to understand the difference in the shapes of the heat capacity curves of pure p-H$_2$ and pure o-H$_2$ shown in Fig. 7.3 (the ortho- and para-states of hydrogen are discussed in the next chapter). Let us consider first the three lowest rotational levels that can be occupied by p-H$_2$; they are $J = 0, J = 2$ and $J = 4$. The spacings between these levels are $6B(J = 0 \to 2)$ and $20B(J = 0 \to 4)$ as calculated from the rigid rotor equation

$$E_{\text{rot}} = BJ(J+1) \,.$$

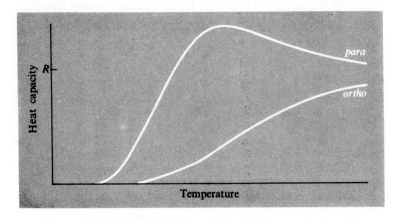

Fig. 7.3. Rotational heat capacities of o-H_2 and p-H_2.

Because of the high energy of the $J \geqslant 4$ states we can make an approximate treatment of p-H_2 at low temperatures as a two level, Schottky system, the levels being $J = 0$ and $J = 2$. Now the ratio of the degeneracies of these two level, $(p_{J=2})/(p_{J=0})$ is $(2 \times 2 + 1)/(2 \times 0 + 1) = 5/1$. As Table 7.1 and Fig. 7.2 show we should then expect a heat capacity which initially rises rapidly with increasing temperature to a maximum of about $1.44R$; i.e. to well above the classical value of R for a diatomic molecule. This behaviour is just what is observed. As the temperature is raised further the heat capacity falls, not to zero as in a true Schottky system, but to the classical limit as the higher energy levels begin to become populated and make their contribution to the rotational heat capcity.

By contrast, the lowest two levels of o-H_2 are $J = 1$ and $J = 3$, for which the degeneracy ratio is $(2 \times 3 + 1)/(2 \times 1 + 1) = 7/3$. On the Schottky model the maximum heat capacity for this ratio is less than R and there is no tendency for the energy levels to become populated other than in a regular sequence. C_{rot} of o-H_2 thus rises smoothly to the high-temperature value, without passing through a maximum.

The Importance of kT

The value of kT when compared with the value of ε is of great importance in deciding whether the temperature in a particular situation is 'high' or not. Although such comparisons between ε and kT have to be made in a wide variety of problems, statistical thermodynamics gives us no idea of what the actual energy quantities are likely to be. It is found that they may be equivalent to a temperature anywhere between the 60,000 K characteristic of electrons in a metal, to the 10^{-5} K characteristic of the interaction of nuclear spins with one another.

Some theoretical treatments of important situations found in chemical systems require that the energy of interaction of particles should be much less than kT. Here are two such examples.

(i) In Debye-Hückel theory.

The Debye-Hückel theory seeks to explain the non-ideal behaviour of ionic solutions in terms of purely electrostatic interactions between the ions. In this theory it is important that the potential energy of an ion, which is given by $q\psi$ (where q is the charge and ψ the electric potential) should be much less than kT. The reason for this is an example of a mathematical approximation which is frequently made in physical chemistry and which arises as follows. We know from the Boltzmann distribution law that the local concentration of ions, N_i' which are in a small volume of the solution near a particular ion, is

$$N_i' = N_i e^{-\varepsilon/kT}$$

where N_i is the bulk concentration and ε is the electrical potential energy of the ions. Now the density of charge, σ, in this small volume is the sum of the number of ions each multiplied by its charge, i.e.

$$\sigma = \sum_i N_i' q_i$$

$$= \sum_i q_i N_i e^{-\varepsilon/kT} .$$

The exponential term can now be expanded using

$$e^{-\varepsilon/kT} = 1 - \frac{\varepsilon}{kT} + \frac{1}{2}\left(\frac{\varepsilon}{kT}\right)^2 + \cdots$$

In order to obtain an expression for σ which is mathematically tractable and which will ultimately allow the derivation of the Debye-Hückel formula for the dependence of activity coefficient on concentration, it is necessary to cut off this expansion after the linear term. This is a valid approximation only when $\varepsilon \ll kT$. The physical significance of this approximation can be seen by substituting for ε_i its value $q_i\psi$. The condition that the electrical potential energy $q_i\psi$ is much less than the thermal energy, kT, will be most closely fullfilled when q_i and ψ are made as small as possible. Experimentally this is achieved by working with uni-univalent electrolytes at low concentrations in solvents of high dielectric constant, and it is exactly with these solutions that the Debye-Hückel theory has been most successful.

(ii) In the study of dipole moments.

The interaction energy ε_p between a dipole μ and a field F is:

$$\varepsilon_p = -\mu F \cos\theta$$

where θ is the angle between μ and F as in Fig. 7.4. The field tends to align dipoles in the direction of the electric field, while their thermal energy kT tends to keep them randomly oriented.

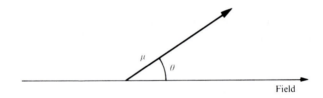

Fig. 7.4. Dipole moment in an electric field.

There is a distribution of the dipole moments amongst all the possible directions. When the moment is against the field direction the molecule has a higher energy and when the moment is along the field it has a lower energy than in a field-free environment. We can use the distribution law to deduce a value for the average moment $\bar{m}$ of a molecule along the direction of the field. The result is

$$\bar{m} = \mu L(u)$$

where $u = \mu F/kT$ and $L(u)$ is the *Langevin function*:

$$L(u) = \frac{e^u + e^{-u}}{e^u - e^{-u}}.$$

Provided that $\mu F \ll kT$ the Langevin function, which can be expanded as the power series,

$$L(u) = u/3 - u^3/45\ldots$$

can be approximated by

$$L(u) = u/3.$$

Then so long as the thermal agitation is considerable it is right to use this approximation to obtain the expression for the orientation polarization

$$P_0 = \frac{4}{3}\pi N \frac{\mu^2}{3kT}.$$

Measurements of the dielectric constant allow the dipole moment to be calculated. However, if kT is not much greater than μF the molecular dipole moments become aligned and the material is said to be 'dielectrically saturated'. The dielectric constant then drops enormously and dipole moments can no longer be obtained.

The Production of Low Temperatures

To achieve cooling we have to reduce the entropy of a system. We therefore require some property, X, other than temperature, on which the entropy depends, i.e.

$$S = f(X, T).$$

Thus we can produce changes in X by varying T and conversely (the case in which we are interested) changes in T by varying X. We can depict this diagrammatically as in Fig. 7.5, which shows the variation of S with T for two particular values of X.

To cool the system we start at A and alter X isothermally from X_1 to X_2. The system loses entropy $(\partial S/\partial X)_T \Delta X$ and an equivalent amount of heat $T_1 \Delta S$. We then restore X to its original value by a reversible adiabatic process from B to C. During this process the entropy is constant, with result that the temperature drops from T_1 to T_2.

In common practice X is either the pressure of a gas or the magnetic moment of a paramagnetic salt. The pressure can obviously be altered at will and the magnetic entropy associated with electron spin can be reduced by applying a strong magnetic field to the sample.

The third law of thermodynamics requires that for both X_1 and X_2 the entropy shall go to zero at absolute zero. This means that if we carry out a finite number of cooling processes we shall still be a finite distance away from absolute zero. This result has been made the basis of the alternative formulation of the third law: 'It is impossible to reach absolute zero in a

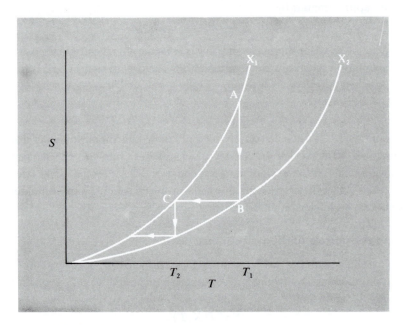

Fig. 7.5. Entropy changes in successive cooling processes.

finite number of steps'. This formulation is not, however, particularly useful in dealing with the properties of substances under ordinary conditions.

Magnetic Entropy

Before we discuss the use of magnetic effects as a means of producing very low temperatures, a few words must be said about magnetic entropy itself. From the familiar Boltzmann relationship we have

$$S = k \ln W .$$

Thus

$$S_{\mathrm{mag}} = k \ln W_{\mathrm{mag}}$$

where W_{mag} can be identified with the number of orientations that the paramagnetic ion can take up in an applied magnetic field. We know from quantum theory that this number is $(2J+1)$, where J is the total electronic angular momentum of the ion; e.g. for $J = 2$ there are five possible orientations in the field B_0, each of which is denoted by M_J. Using the vector model of the atom we can depict the possibilities as in Fig. 7.6. (Note that this J should not be confused with the rotational quantum number, which is also usually given the symbol J.) The magnetic J arises from the interactions between both the spin and the orbital momenta of all the unpaired electrons in the atom and is written as a subscript in the symbol used to describe the electronic state of an atom e.g. for sodium with one electron ($S = 1/2$) in a p-orbital ($L = 1$) J can be 3/2 i.e. ($L + S$) or 1/2 i.e. ($L - S$). There are then two closely spaced atomic energy levels $^2P_{\frac{3}{2}}$ and $^2P_{\frac{1}{2}}$ (the superscript 2 is called the multiplicity of the state and is defined as $2S + 1$ i.e. $(2 \times 1/2 + 1) = 2$).

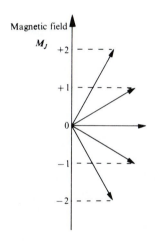

Fig. 7.6. Orientations of a magnetic moment in a magnetic field.

In the absence of a magnetic field the $(2J+1)$ states are of equal energy (degenerate) and any particular ion has an equal probability of being in any of the states. For N ions therefore

$$W_{\text{mag}} = (2J + 1)^N$$

$$S_{\text{mag}} = k \ln W_{\text{mag}} = R \ln(2J + 1).$$

When a magnetic field is applied the degeneracy is lifted and a series of energy levels is produced because each of the orientations illustrated in Fig. 7.6 is of different energy. The magnetic interaction energy, ε_{mag}, is given by

$$\varepsilon_{\mathrm{mag}} = -M_J g_J \mu_B B_0$$

where M_J is shown in Fig. 7.6, g_J is a constant for a particular ion (the *Landé factor*), μ_B is a fundamental constant (the *Bohr magneton*), and B_0 is the applied field.

Following the third law, the zero-field magnetic entropy of $R\ln(2J+1)$ should disappear at low temperatures; and indeed below a temperature characteristic of a particular compound there is often magnetic ordering, in one of two ways. Schematically, these can be represented thus:

↑↑↑↑↑

where all the tiny magnets are aligned parallel. This is the ferromagnetic state, and the transition from the disordered (paramagnetic) to ordered (ferromagnetic) arrangement takes place at the *Curie temperature*. Alternatively, antiferromagnets have the components antiparallel in a regular fashion:

↑↓↑↓↑

and the transition point is called the *Néel temperature*. Curie points vary widely; e.g. for iron the Curie point is about 1000 K but for some paramagnetic salts it is < 0.01 K and the latter are used to achieve very low temperatures. At normal temperatures the separation between adjacent magnetic levels $(g\mu_B B_0)$ is much less than kT for any magnetic fields which can conveniently be generated in a laboratory $(B_0 \sim 50\,000$ G, 5 T). The magnetic moments then have their full entropy of $R\ln(2J+1)$. But if the sample is cooled sufficiently and a strong magnetic field is applied the magnetic moments tend to become aligned and the entropy is thereby reduced. This effect is used in producing very low temperatures by 'adiabatic demagnetization'.

Adiabatic Demagnetization

This is the most widely used technique for producing very low temperatures. The first stage is to use a low-boiling liquid; liquid helium-4 is fairly readily available and boils at 4.2 K. If the Dewar containing liquid ^{4}He is pumped so that the liquid helium evaporates rapidly, the latent heat of evaporation reduces the temperature to below 1 K. It is also possible to use ^{3}He which boils at an even lower temperature than ^{4}He.

Next we use the magnetic moment of suitable paramagnetic salt. At 1 K the moments are still randomly orientated but at this low temperature a magnetic field can reduce the randomness. Thus if we plot a diagram of T against S we have Fig. 7.7. To produce the cooling we start at point 1 on the zero field curve of Fig. 7.7. When the field B_0 is applied the magnetic entropy drops to point 2 on the lower curve and heat is given out which is conducted away. The salt is then thermally isolated and the field is switched off. This is the process of 'adiabatic demagnetization'. The salt must now return to the zero field curve under the constraint of constant entropy (because it is isolated). This it does by cooling to the much lower temperature at point 3.

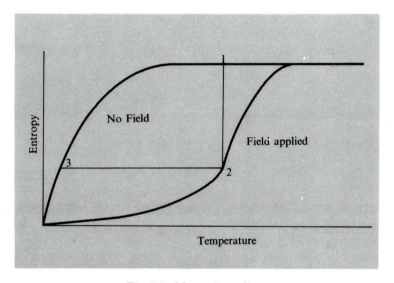

Fig. 7.7. Magnetic cooling.

In this way, temperatures of about 0.003 K can be achieved. Still lower temperatures have been reached using not electronic moments but nuclear moments (which arise because many nuclei have a spin which gives rise to a nuclear magnetic moment). The starting point for a nuclear cooling process is ~ 0.01 K and temperatures of ~ 0.0000005 K have been attained.

Liquid Helium

Helium, which exists in two isotopic forms ^{4}He and ^{3}He, is unusual in being liquid to very low temperatures, and indeed its low-temperature behaviour is fascinating. Liquids are usually associated with more disorder than solids but He is exceptional in this respect.

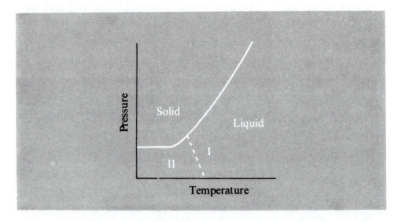

Fig. 7.8. Phase diagram of liquid helium ^{4}He.

Let us first consider ^{4}He. A considerable pressure (25 atm) has to be applied to solidify the liquid. The phase diagram, Fig. 7.8, shows that the slope of the melting curve tends to zero as $T \to 0$. Then, using the Clausius-Clapeyron relationship

$$\frac{\mathrm{d}p}{\mathrm{d}T} = \frac{\Delta S}{\Delta V}$$

we find that, since $\mathrm{d}p/\mathrm{d}T = 0$ and the volume change is observed to be constant, $\Delta S = 0$, i.e. $S_l = S_s$. We conclude then that at very low temperatures the liquid is extremely ordered. Furthermore, there is virtually no latent heat of fusion since $L = T\Delta S \simeq 0$. These novel observations are peculiar to helium at temperatures below 2.17 K. At that temperature there is a transition from the normal behaviour of liquid He-I to the unique behaviour of the low temperature form, liquid He-II. The transition is marked by an anomaly in the heat-capacity curve, called a *lambda point* (see Chapter 10). He-II has other unusual properties and in some ways behaves rather like a solid. For example:

(i) its heat capactiy follows the Debye T^3 law

(ii) its thermal conductivity is similar to a solid's and is proportional to the diameter of the specimen.

Perhaps the most remarkable property of liquid He-II is its superfluidity i.e. its viscosity becomes vanishingly small. This unique property of ^{4}He at very low temperatures is a consequence of quantum effects. The fact that ^{4}He obeys Bose-Einstein statistics means that there is no limit to the number of particles that can occupy any particular energy level. (We discussed this in Chapter 2.) At very low temperatures, therefore, all the particles of ^{4}He can congregate in the lowest level and they tend to do so over rather a narrow temperature range near 2 K. It is this process, referred to as Bose-Einstein condensation, which causes the heat capacity anomaly at 2.17 K. When the particles have undergone this condensation the properties of the system change radically and the unique features mentioned above are observed.

The isotope ^{3}He is rare in nature but can be produced artificially by nuclear decay processes. It has the same electronic structure as ^{4}He but as a result of a lighter mass it has a lower boiling point than ^{4}He. Like liquid ^{4}He, liquid ^{3}He requires a considerable pressure to solidify it. The melting point curve of ^{3}He has a curious shape as Fig. 7.9 shows. Below about 0.3 K the slope of the curve becomes negative i.e. $dp/dT < 0$. If we use the Clausius-Clapeyron equation and the observation that ΔV for melting is approximately constant we conclude the ΔS is negative i.e. the liquid is more ordered than the solid. At these very low temperatures the entropy of ^{3}He is largely due to the random orientation of the nuclear spins ($I = 1/2$ for ^{3}He). The unexpected result of a negative ΔS implies that the nuclei are more aligned in the liquid than in the solid, a conclusion which is supported by magnetic susceptibilitiy measurements. The reason for the difference may be that the forces between the nuclei which tend to cause ordering ('exchange forces') are less in the solid than in the liquid, where collisions allow a greater interaction. The unusual properties of ^{4}He which result from Bose-Einstein condensation were not expected for ^{3}He, since this isotope must obey Fermi-Dirac statistics. However, it has been discovered very recently that at very low temperatures even ^{3}He will act as a superfluid. This is probably due to the ^{3}He nuclei moving in pairs when they act as bosons and can become superfluid by a mechanism analogous to that which gives rise to superconductivity, which we shall discuss a little later.

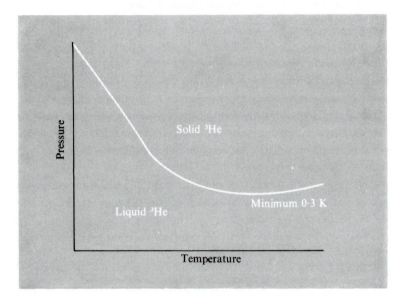

Fig. 7.9. Melting point curve of ^{3}He.

The Uses of He in Producing Low Temperatures

Apart from using liquid He just as a refrigeration liquid, there are two other ways in which the properties of the substance can be used.

The first is *dilution refrigeration*. Below 0.8 K the two isotopes separate spontaneously. This is an example of the reduction in the entropy of mixing, which when thermodynamic equilibrium is maintained, tends to zero at absolute zero. In fact the two forms do not separate completely; the lower phase is the heavier ^{4}He with a little ^{3}He dissolved in it, and floating on top is a ^{3}He layer which is practically pure. At these low temperatures the two isotopes have quite different properties. ^{4}He with zero nuclear spin has undergone Bose-Einstein condensation and has a heat capacity close to zero. On the other hand the nuclear spins of ^{3}He remain disordered down to about 0.1 K and this isotope has a considerable heat capacity between 0.1 and 0.8 K. We can think of the two phases as a ^{4}He layer which is dilute ^{3}He in an inert atmosphere and a ^{3}He layer which is concentrated—rather like a vapour and condensed liquid but with the dilute (vapour-like) phase underneath. The parallel extends further, because just as when a liquid evaporates, the passage of ^{3}He from the upper layer to the lower induces cooling (Fig. 7.10). This process forms the basis of the helium dilution refrigerator. Its main advantage is that it can maintain very low temperatures for a long time.

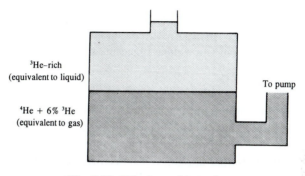

^{3}He–rich
(equivalent to liquid)

^{4}He + 6% ^{3}He
(equivalent to gas)

To pump

Fig. 7.10. Dilution refrigeration.

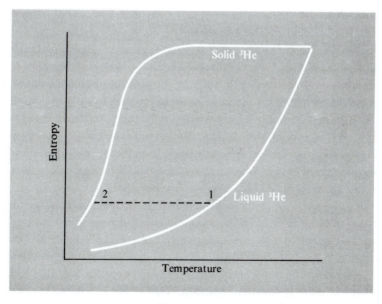

Fig. 7.11. Low-temperature entropy of ^{3}He.

Liquid is in equilibrium with its vapour. Solid is under pressure.

The other method of cooling which utilizes the unique properties of ^{3}He at low temperatures was proposed by Pomeranchuk. His idea depends on the fact, already mentioned, that liquid ^{3}He is more ordered than solid ^{3}He below 0.318 K. The entropy-temperature diagram is shown in Fig. 7.11. The liquid is in equilibrium with its vapour and the solid is under pressure. If we take some liquid ^{3}He, say at point 1, and apply pressure adiabatically we can solidify to point 2 with consequent cooling. Only technical difficulties have hindered the wide application of this method.

Superconductivity

Perhaps of all the fascinating things which happen at very low temperatures the most startling is superconductivity. When certain metallic substances are cooled sufficiently a current once started will persist even after the applied potential difference has been removed. There is no resistance and we have a sort of perpetual motion. The possible commercial uses are obvious and are already being exploited. These include loss-less power transmission, very powerful electromagnets, motors, and computer memory elements.

The explanation of the phenomenon was given by Bardeen, Cooper and Schreiffer (BCS theory) following some ideas originally presented by Fröhlich and London. In a metallic conductor we have free electrons, which as fermions, obey the Pauli principle. In the absence of an applied voltage, the same number of electrons move in each direction in the wire. A current may be induced by applying a potential difference but it dies away when the p.d. is switched off and the electrons resume their lowest-energy state.

In superconductors the electrons move in pairs and each pair can have an overall spin of 0 or 1. That is to say, the pair behaves as a boson rather than two separate fermions. Now, as we have seen, bosons can all have the same energy state or momentum state. It is therefore possible for the electrons to move in pairs in the same direction with the same momentum. This is equivalent to a zero resistance and, once started, they will continue indefinitely if the state is stabilized. Thus the problem of understanding the basis of superconductivity becomes one of understanding why the electrons can move as pairs at low temperatures. The reason can be appreciated by means of the analogy of two marbles rolling on a drum. If the drum is shaken violently the two marbles will move completely independently of each other and quite randomly. If however, we just tilt the drum gently to and fro the marbles will move as a pair, each in the small trough in the drum created by the other (Fig. 7.12(a)). The motion of a pair of electrons in a fairly rigid lattice of positive charges can be similar (Fig. 7.12(b)). If an electron moves between two positive charges they come together slightly, making a region of increased potential to attract the second electron. This is repeated at the next site and so the two electrons can move as a pair. If we increase the temperature, the lattice vibrations become violent enough to separate the pair and this is why superconductivity is a low-temperature phenomenon.

Fig. 7.12.   (a) Movement of marbles on a drum illustrating the formation of Cooper pairs.

Fig. 7.12.   (b) Movement of electrons in pairs through a lattice, giving rise to superconductivity.

High Temperature Superconductors

A major research goal has been to find materials which would become superconducting at higher and higher temperatures, since if constrained to liquid helium temperatures, any exploitation would be limited. After years of slow progress this line of research took a huge step forward in the late 1980s, when it was discovered that some complex inorganic materials with layered structures were capable of raising the crucial temperature by tens of degrees. So-called high-critical-temperature superconductivity has now been found in a rich variety of materials including Fullerenes, $(Ba,K)BiO_3$ and layered perovskites with layers of LuC alternating with nickel boride tetrahedra. One example of a superconducting structure is shown in Fig. 7.13. There is no doubt that this will remain an area of enormous research activity.

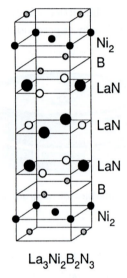

Fig. 7.13. The layered structure of the superconducting materials $La_3Ni_2B_2N_3$.

Chapter 8

THERMODYNAMICS AND SPECTROSCOPY

The observation of the intensities of spectral lines provides the most direct indication of the variation in the populations of energy levels in molecules which is a central theme of this book. The intensity is governed both by quantum-mechanical considerations and by the statistical thermodynamic distribution of particles.

Spectroscopic Selection Rules

The ability of electromagnetic radiation to excite an atom or molecule from one energy level to another depends on the wave-functions describing the two states. For the normal spectroscopic transitions observed at any frequency between the ultraviolet and microwave regions of the spectrum (but excluding magnetic resonance experiments) it is the electric dipole component of the radiation that causes the transition. It is province of wave-mechanics rather than of thermodynamics to predict whether in any particular case this oscillating dipole can indeed cause a transition to occur. When the predicted probability of the transition is non-zero, the transition is described as 'allowed' whereas, when the probability is zero, the transition is said to be 'forbidden'. The conditions under which a transition is allowed form the basis of the various spectroscopic selection rules. There are, however, various degrees of 'forbiddenness' among selection rules and many so-called forbidden transitions can be observed. This can happen because the theoretical treatment may relate to an idealized situation from which the real system departs to some extent. Thus, for example, the selection rule for vibrational transitions in molecules is $\Delta v = \pm 1$, where v is the vibrational quantum number. However, this rule is arrived at by treating the molecule as a simple harmonic oscillator and the deviation of potential energy curves from the parabolic shape characteristic of simple harmonic motion allows other transitions, in which v changes by 2 or 3, to be observed, though with a much lower intensity than the main transition.

The absorption of light in the visible region, which is associated with electronic transitions within molecules, provides a good example of the spread of transition probabilities which can be observed. The relative probability can conveniently be described by the 'extinction coefficient' at an absorption maximum. This coefficient is defined by the Lambert-Beer Law for monochromatic radiation

$$I = I_0 \, e^{-\varepsilon c d}$$

where ε is the extinction coefficient, c is the concentration of absorbing species and d is the thickness of the cell. For the intense colour of a dye, ε is $\sim 10^5$ dm^3 mol^{-1} cm^{-1}, whereas for the very pale pink colour of Mn^{2+} in aqueous solution $\varepsilon \sim 10^{-1}$ dm^3 mol^{-1} cm^{-1}. The sensitivity of light-recording devices, including the human eye, thus allows a wide range of electronic transitions to be observed.

Intensities of Lines in Vibrational-Rotation Spectra

Thermodynamic considerations give no information about the selection rules applicable for any particular transition within a molecule. However, once it has been shown that a transition can be observed and the selection rules have been established, these considerations can play an important role in discussing the fine structure which is so frequently observed in molecular spectra. This is possible because the relative intensities of the fine-structure lines are determined largely by the populations of the energy levels from which the transitions originate. These are in turn determined by the distribution law.

To illustrate how this works out in practice let us consider in some detail the vibrational-rotation spectrum of HCl (gas) which is shown in Fig. 8.1. This is an absorption spectrum, in which the selection rule for the vibrational quantum number v is $\Delta v = +1$ ($\Delta v = -1$ is allowed, but corresponds to the emission of radiation from an upper vibrational level

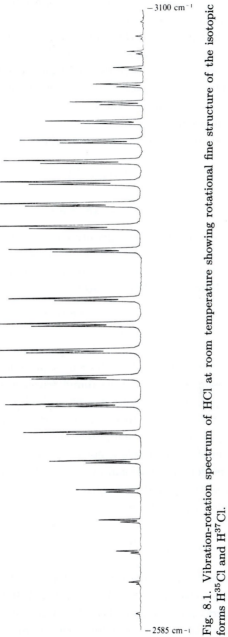

Fig. 8.1. Vibration-rotation spectrum of HCl at room temperature showing rotational fine structure of the isotopic forms H^{35}Cl and H^{37}Cl.

and is not observed at room temperature because the population of the upper level is very small). For a diatomic molecule there is an additional selection rule which requires that the rotational quantum number, J, must also change during the transition, according to the selection rule $\Delta J = \pm 1$. Non-linear molecules can undergo vibrational transitions without a change of rotational quantum number and the selection rule is then modified to $\Delta J = 0, \pm 1$. This also applies to the bending (perpendicular) modes of linear polyatomic molecules.

The energy levels of a simple harmonic oscillator are:

$$E_v = h\nu\left(v + \frac{1}{2}\right)$$

$$E_{v+1} - E_\nu = \Delta E = h\nu .$$

The frequency, ν, which defines the separation of the energy levels is about 2900 cm^{-1} for HCl. This is equivalent to an energy of 34.7 kJ mol^{-1}. If we now calculate the room-temperature ratio of molecules in $v = 1$ to those in $v = 0$ from the Boltzmann distribution law:

$$n_{v=1} = n_{v=0} \, e^{-34\,700/298\,R}$$

we find that $n_{v=1} \approx 10^{-6} \times n_{v=0}$. This very low population of excited vibrational states means that when the gas is at room temperature we need only concern ourselves with the ground vibrational level, though if the gas is heated sufficient molecules can be excited to the $v = 1$ level to allow the transition from $v = 1$ to $v = 2$ to be observed. This transition comes at a slightly different frequency from the $v = 0$ to $v = 1$ transition because the motion of the molecule is not truly simple harmonic.

Now let us consider the rotational motion of the molecule. As we saw on p. 40, solution of the Schrödinger equation for a rigid rotating diatomic molecule gives the energy levels as

$$E_J = \frac{h^2}{8\pi^2 I} \cdot J(J+1) = BJ(J+1) .$$

For HCl, B is about 10 cm^{-1} which is 119 J mol^{-1}. At room temperature B_{HCl} is considerably less than kT (which is about 200 cm^{-1}) so that several rotational energy levels are significantly populated. In calculating the populations of the rotational levels a further conclusion from the solution of the wave-equation must be taken into account. This is that for any rotational quantum number J there are $(2J + 1)$ solutions of equal energy. In calculating the populations, therefore, a statistical weight factor $(2J + 1)$ must be included, and the distribution law becomes

$$n_J = \frac{N}{q}(2J + 1)\exp(-BJ(J + 1)/kT).$$

Now let us consider how n_J calculated from the above equation varies as J is increased. For a system at constant temperature and volume, N and q are constant and the populations of the energy levels are therefore governed by the balance between the linear pre-exponential term $(2J + 1)$ which is tending to increase the populations of the higher levels and the exponential term $\exp(-BJ(J+1)/kT)$ which is tending to a rapid fall-off in population as J is increased. The variation in population of the rotational energy levels of HCl at room temperature is shown in Fig. 8.2 and may be seen to be quite different from the vibrational case. At room temperature the balance between these two opposing factors leads to a maximum in n_J at about $J = 3$. Above this value the exponential term becomes increasingly

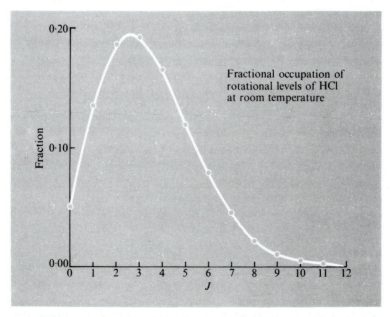

Fig. 8.2. Fractional occupation of rotational levels of HCl at room temperature.

dominant so that when J is greater than about 10 the populations are negligible. A comparison of these predicted populations with the experimentally observed i.r. peak heights shows that both vary in a similar way. Bearing in mind that peak heights are only an approximate measure of the populations this agreement shows that the thermodynamic discussion does give a satisfactory interpretation of the relative intensities of the spectral lines.

The discussion can be taken a stage further in order to predict how the spectrum will change if the sample temperature is changed. Suppose first that the gas is cooled below room temperature. The molecules in upper rotational levels will then drop back into the lower levels with the result that the intensities of the higher rotational quantum lines diminish. Both branches of the spectrum then appear to be more sharply peaked at the lower values of J, and the whole spectrum will look more compressed.

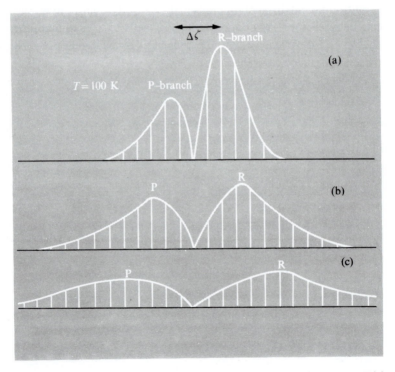

Fig. 8.3. Calculated variation of i.r. spectrum with temperature $T(c) > T(b) > T(a)$.

Conversely if the gas is heated more of the upper levels will be populated at the expense of the lower levels. The spectrum will then cover a greater range and the two maxima will be less pronounced, and will be further apart. The separation between the two maxima, $\Delta\zeta$, thus depends on the temperature (Fig. 8.3) and it also depends on the constant B. This dependence is of the form

$$\Delta\zeta \propto \sqrt{(BT)}.$$

In cases where the fine structure is not resolved, this formula has been used to obtain approximate values for B. If on the other hand we can observe the spectrum of a molecule of known structure, i.e. of known B, we can determine the temperature of the gas. This is one of the ways in which astral temperatures can be measured, and was employed in determinations of the temperature of the atmosphere of Venus.

Ortho- and *Para-*Hydrogen

For many of the discussions of the thermodynamic properties of molecular systems the existence of nuclear spin can be ignored without introducing error. However, as we saw in the last chapter, one important and interesting case in which nuclear spin must be taken into account is hydrogen. The special thermodynamic properties of hydrogen which can be attributed to nuclear spin arise because it is a homonuclear diatomic molecule with a particularly low moment of inertia combined with a low boiling point. Nuclear spin also has an effect on the intensities of the spectral lines of all homonuclear diatomic molecules and, as in the discussion of the vibration-rotational lines, it is sensible to discuss the two phenomena together.

The discussion starts from a general consideration of the symmetry properties of the wave-functions which describe any system of identical fundamental particles. We have discussed this earlier in Chapter 2, but it is of such central importance that it merits some restatement in a more rigorous fashion. We are not concerned with the mathematical form of these equations but only with what happens to the sign of the wave-function when two identical particles are interchanged. Now a wave-function, Ψ, which is a solution of the wave equation for the system under discussion is a purely mathematical entity and has no direct physical significance. The physically significant quantity is Ψ^2 (or more strictly $\Psi\Psi^*$), which is often described as the 'probability' of the system; thus, for example, when Ψ is a solution of the Schrödinger equation for the hydrogen atom, $\Psi^2 \, d\tau$ is the probability that the single electron will be found in a small volume element $d\tau$. Now let us interchange any two of the identical particles. The resulting situation must be physically indistinguishable from the original; that is to say, Ψ^2 must be unchanged by the interchange of identical particles. However, there are two possibilities for Ψ which leave Ψ^2 unchanged. Either Ψ is itself unchanged, in which case it is 'symmetric', or Ψ changes to $-\Psi$ and is described as 'antisymmetric'. Certain generalizations have emerged from experience (the existence of fermions and bosons mentioned in Chapter 2) and these observations are supported by recent developments in field theory. Thus, if electrons, protons and neutrons (all of which are fermions with half-integral spin) are regarded as 'building blocks' for nuclei and atoms it has been found that when the particles under consideration are made up of an odd number of building blocks the wave-function changes sign on interchange of particles, whereas when they are made up of an even number of building blocks the wave-function is unchanged by interchange of particles. It follows that, for example, all the wave-functions for any system of hydrogen nuclei or any system of electrons are antisymmetric, whilst for a system of deuterium nuclei the wave-functions are unchanged by the exchange of particles.

These symmetry properties with respect to interchange are maintained when the particles become parts of atoms or molecules. Probably the most familiar example of this is the requirement that the total wave-functions of electrons in atoms shall be antisymmetric. As we have seen earlier, this is one way of stating the Pauli Exclusion Principle, which is more commonly expressed as the requirement that no two electrons in an atom shall have all four quantum numbers the same.

When we come to consider the effect of the requirement that the total wave-function for the hydrogen molecule must be antisymmetric with respect to interchange of the nuclei, it is helpful to make use of the Born-Oppenheimer approximation. This allows the internal energy of a molecule to be divided into essentially independent components which are due either to electronic interactions or to the movement of the nuclei. The latter part can itself be divided into contributions arising from the vibrational and rotational motion of the molecule and to nuclear spin. The total energy can therefore be written as

$$E_{\text{total}} = E_{\text{el}} + E_{\text{vib}} + E_{\text{rot}} + E_{\text{ns}}$$

(where the subscript ns stands for nuclear spin) from which it follows that

$$\Psi_{\text{total}} = \Psi_{\text{el}} \times \Psi_{\text{vib}} \times \Psi_{\text{rot}} \times \Psi_{\text{ns}}.$$

The symmetry of Ψ_{total} is thus determined by the product of the symmetries of the individual wave-functions. The correct symmetry is arrived at if symmetric is counted as $(+)$ and antisymmetric is counted as $(-)$ and the normal rules of multiplication are applied. We shall not need, therefore, to know the explicit mathematical form of the various wave-functions but only their symmetries. All combinations of wave-functions are allowed in principle provided only that their product is antisymmetric, i.e. $(-)$, if fermions are exchanged and symmetric, i.e. $(+)$, if bosons are interchanged. Now let us consider the symmetry properties of the various wave-functions in turn.

For a diatomic molecule the vibrational wave-function depends only on the separation of the nuclei and not on their disposition in space. It therefore remains unaffected by an interchange of nuclei and is $(+)$. For the simple (close-shell) molecules we consider the contribution of Ψ_{el} can most readily be understood if the process of nuclear exchange is divided into two consecutive operations. In the first of these the coordinates of both the electrons and the nuclei are inverted at the origin, a process which is acheived by replacing the spatial cordinates (x, y, z) by $(-x, -y, -z)$ wherever they occur in the wave-functions. In the second operation the coordinates of the electrons alone are inverted, thus restoring the electronic wave-function to its original condition while leaving the nuclei interchanged. For hydrogen in its ground state the electronic wave-function is symmetric with respect to these successive operations and is $(+)$. We are now left with the situation that the symmetry of Ψ_{total} is determined by the products of the symmetries of Ψ_{rot} and Ψ_{ns}. The solution of the Schrödinger wave-equation for a rigid rotor, to which the hydrogen molecule closely approximates, gives wave-functions which are alternately symmetric and antisymmetric with respect to the first of the two operations we have carried out for exchanging nuclei. Fig. 8.4 shows these wave-functions for $J = 0, 1$, and 2.

The wave-function is symmetric for even values of J and antisymmetric for odd values of J. This leaves the symmetry of Ψ_{ns} to be determined. Now the hydrogen nucleus has a nuclear spin, $I = 1/2$, so that two nuclear spin states are possible, which we shall call α and β. Because the nuclei are identical, the nuclear spin wave-functions must themselves be either

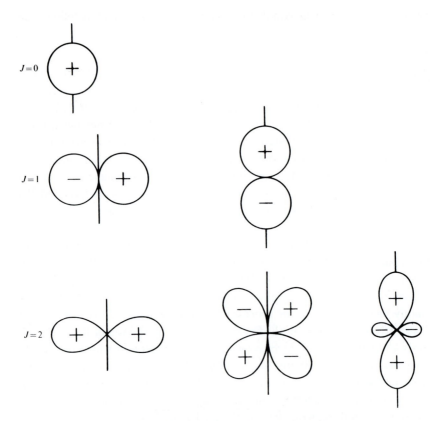

Fig. 8.4. Rotational wave-functions for a rigid rotor.

symmetric or antisymmetric with respect to interchange of the nuclei. The allowed nuclear spin states are therefore

$$\Psi_{ns} = \alpha\alpha$$

or

$$\Psi_{ns} = \beta\beta$$

or

$$\Psi_{ns} = \alpha\beta + \beta\alpha$$

or

$$\Psi_{ns} = \alpha\beta - \beta\alpha$$

Of these the first three are symmetric and the last one is antisymmetric. There are thus three times as many symmetric nuclear spin wave-functions as there are antisymmetric functions. This 3:1 ratio, which is of great importance in determining the intensities of spectral lines, can also be deduced as follows by using a vector model for the nuclear angular momentum. Two nuclei with $I = 1/2$ can combine their angular momenta vectorially to give a total nuclear spin vector, T, of either 1 or 0. If the nuclear spin vector is represented by an arrow, these two possible nuclear spin orientations can be written schematically as ($\uparrow\uparrow$) for $T = 1$ and ($\uparrow\downarrow$) for $T = 0$. As for any angular momentum vector, the statistical weight of either of the nuclear spin configurations is $(2T+1)$, which again yields a ratio of 3:1. By convention the state with the greater statistical weight is called the *ortho* state; the other is *para*.

Having identified the symmetry properties of the various wave-functions, for the hydrogen molecule in its ground state, we are now in a position to find the combinations which make Ψ_{total} anti-symmetric. Since Ψ_{el} and Ψ_{vib} are both $(+)$, the overall symmetry is determined by the product of Ψ_{rot} and Ψ_{ns} so that one but not both of these must be antisymmetric. Therefore when Ψ_{rot} is symmetric and J has even values, Ψ_{ns} must be anti-symmetric $(T = 0)$, and when Ψ_{rot} is antisymmetric and J has odd values, Ψ_{ns} must be symmetric $(T = 1)$. We can now write Ψ_{total} in either of two ways:

$$para \qquad \Psi_{total} = \Psi_{el} \times \Psi_{vib} \times \Psi_{rot} \times \Psi_{ns}$$

$$J = even \quad T = 0$$

$$(+) \quad (+) \quad (+) \quad (-).$$

$$ortho \qquad \Psi_{total} = \Psi_{el} \times \Psi_{vib} \times \Psi_{rot} \times \Psi_{ns}$$

$$J = odd \quad T = 1$$

$$(+) \quad (+) \quad (-) \quad (+).$$

It is thus seen that molecules which are in the *ortho*-state $(T = 1)$ can only occupy odd rotational levels whereas molecules in the *para*-state $(T = 0)$ can only occupy the even rotational levels. Now the nuclei interact weakly with external influences because of their small magnetic moments so that it is difficult to effect the reversal of one of the nuclear spins which is required to convert the molecule from an *ortho* to a *para* state or *vice versa*. This conclusion is embodied in a spectroscopic selection rule for homonuclear diatomics which states that transitions are possible only between rotational states of the same symmetry. Thus when the molecule changes its rotational quantum number, ΔJ must be ± 2. When hydrogen is cooled to its boiling point the molecules therefore fall either to $J = 0$ if they were originally in even J states or to $J = 1$ if they were in odd J states. The difficulty of interconverting *ortho* and *para* states requires the hydrogen to be maintained in its liquid form for a considerable time before the true thermodynamic equilibrium is achieved, in which almost all the molecules are in $J = 0$ and nearly pure *para*-H_2 is formed. This low-temperature equilibration can be hastened by adding a suitable catalyst, such as highly active charcoal or a paramagnetic salt. The intense local magnetic fields which are generated by such materials are able to introduce sufficient perturbations at the nuclei to reverse the nuclear spins and allow the *ortho*-molecules in $J = 1$ to become *para*-molecules and fall to $J = 0$. If this *para*-H_2 is now allowed to warm up to room temperature in an inactive vessel it remains as the *para* form for many hours. In this state it has been a valuable aid to the investigation of some reactions involving hydrogen, because the thermodynamic properties of *para*-H_2 are sufficiently different from the properties of the equilibrium mixture to allow an easy and quantitative determination to be made of the extent to which the *para* has been converted. The thermal conductivity is one convenient measure of the composition and has been used, for example, in investigations of the mechanism of the reaction of hydrogen with various metal surfaces.

The difficulty of inter-converting *ortho* and *para* hydrogen means that under some circumstances the gas must be treated as a mixture of two independent components for which separate partition functions must be written. We shall now discuss some consequences of this separation. The first interesting aspect of the mixture is its composition. We have already seen that at low temperatures the equilibrium composition consists of the essentially pure *para* form. We can calculate the high temperature equilibrium composition from the partition functions for the *ortho* and *para* states. At room temperature the molecules are in the ground electronic and vibrational states so that we need to consider only the rotational partition function. The energy of any rotational level is $BJ(J + 1)$ and its rotational statistical weight is $(2J+1)$. In addition each level has a nuclear statical weight of $(2T+1)$ where $T=0$ for *para* and $T=1$ for *ortho*. Thus we may write:

$$q_{para} = (2T+1)_{T=0} \sum_{J=0,2,4...} (2J+1)\,e^{-BJ(J+1)/kT}$$

$$q_{ortho} = (2T+1)_{T=1} \sum_{J=1,3,5...} (2J+1)\,e^{-BJ(J+1)/kT}$$

Now the relative numbers of *ortho* to *para* molecules will be in the ratio of q_{ortho} to q_{para}, i.e.

$$\frac{n_{ortho}}{n_{para}} = \frac{q_{ortho}}{q_{para}}.$$

At high temperatures the summations in the equations for q_{para} and q_{ortho} become closely similar so that the ratio of *ortho* to *para* is determined by the nuclear statistical weight factor $(2T+1)$. Thus at temperatures above about 300 K

$$\frac{n_{ortho}}{n_{para}} = \frac{(2T+1)_{ortho}}{(2T+1)_{para}} = \frac{3}{1}$$

so that the equilibrium mixture contains 25 per cent of *para*-H_2 and 75 per cent of *ortho*-H_2.

Spectroscopic Properties of *ortho*- and *para*-H_2

Having determined the composition of the gas at the extremes of low and high temperature we may now investigate the effect of the nuclear spin on the intensities of the spectroscopic lines of hydrogen. Two types of transition are of particular interest. The first involves the excitation of a molecule from the ground state to an excited electronic state which will be of opposite electronic symmetry i.e. $(-)$. This transition is often accompanied by a change both in the vibrational and the rotational quantum numbers. It is with the fine-structure of the spectrum which is due to the change of rotational quantum number that we are concerned. The general selection rule for the change of J during the transition is $\Delta J = 0, \pm 1$, though in special cases $\Delta J = 0$ is forbidden. Adjacent lines in the spectrum are therefore expected to originate from adjacent rotational levels. However, when pure *para*-H_2 is used alternate rotational levels (J is odd) have zero population, i.e. every alternate line is missing. As equilibration at room temperature proceeds these missing lines, which are due to the *ortho* molecules in odd rotational levels, appear and when equilibration is complete the 3:1 ratio of *ortho* to *para* is reflected in the spectrum which shows an alternation of intensities between adjacent lines. Those lines which originate at antisymmetric (odd) levels have the greater intensity.

A second interesting spectrum of hydrogen is due to the Raman effect. The vibration of the hydrogen molecule does not involve a change in the dipole moment of the molecule, which is always zero, but it does produce a change in the polarizability of the molecule. The vibrational mode of hydrogen is therefore inactive in the infrared but is active in the Raman spectrum. There is also rotational fine structure, for which the selection rule is $\Delta J = \pm 2$. Once again lines which originate from odd rotational levels have an additional intensity due to the extra statistical weight factor $(2T + 1)$, and there is an alternation of intensities. This effect of the nuclei on the intensities of spectral lines is not limited to hydrogen, but is observed for all homonuclear diatomic molecules. It is important to notice, however, that the above discussion applies only when the nuclei are identical; even different isotopes of the same element do not count as identical, because the interchange of isotopic nuclei does not lead to a physically indistinguishable situation. The symmetry arguments then do not apply. When the nuclei have a spin of 1, e.g. D or ^{14}N, the total wave-function must now be symmetric with respect to interchange of nuclei. This means that symmetric rotational wave-functions are associated with symmetric nuclear spin wave-functions, and antisymmetric rotational wave-functions with anti-symmetric nuclear spin wave-functions. In this case the even rotational levels have the greater statistical weight and the *ortho*-to-*para* ratio is 2 to 1. There is a general relation between the nuclear spin and the alternation of population of the rotational levels which is:

$$\frac{n_{ortho}}{n_{para}} = \frac{I+1}{I} \, .$$

Nuclei with Zero Spin

The formula for the *ortho*-to-*para* ratio has interesting consequences for homonuclear molecules in which the nuclei have zero spin, e.g. $^{16}O_2$, for it implies that the intensity ratio is infinite. This can occur only if alternate lines are completely absent. The reason for these absences, which have been observed experimentally, can be explained by extending our symmetry argument. In order to do so, we shall assume that zero is an even number, an assumption which has been implicit in our earlier discussions. It follows from this that a homonuclear molecule in which the nuclei have zero spin must have a total wave-function which is symmetric with respect to interchange of the nuclei. Thus, as before, the symmetric rotational wave-functions (J is even) are associated with symmetric nuclear wave-function. However, when the antisymmetric rotational wave-functions are considered, a difficulty arises because it is impossible to construct antisymmetric nuclear wave-functions for nuclei of zero spin. This complete absence of antisymmetric wave-functions makes it impossible for molecules to enter the antisymmetric rotational states (J is odd).

There are thus no molecules at all in the odd rotational levels, and clearly no transitions originating from them are possible. Put another way, the probability of a transition to or from odd rotational levels is identically zero and never at any time can molecules enter these states. The absence of alternate lines which is observed experimentally is thus accounted for.

Negative Temperatures

Thermodynamics is essentially concerned with the description of systems at equilibrium. However, there are some situations of interest in which the distribution of the particles, deviates, either temporarily or permanently, from the normal state, due to some external influence. It can be helpful in interpreting the properties of these systems to apply, as far as possible, the consequences of our discussions of the classical distribution law. For simplicity let us consider a system of non-degenerate energy levels, for which we can write

$$n_i = n_0 \exp(-\varepsilon_i/kT).$$

Under all equilibrium conditions $n_i < n_0$.

Now let us suppose that by the application of some external influence we are able to invert the populations of the two levels so that $n_i > n_0$. The equation can be satisfied for $n_i > n_0$ by writing T as a negative quantity. If the system is now allowed to equilibrate with its surroundings, which are at room temperature, the population of the upper level will go down while that of the lower level increases. At some intermediate time it may happen that $n_i = n_0$ in which case $T = \infty$. That is to say, a system which can be described as having a 'negative temperature' returns to a normal temperature by way of infinity and not by way of zero. In fact, a system at a 'negative temperature' is very 'hot' and will give up energy to a system at a normal temperature. This property is the basis of the operation of the laser, which will now be described.

Lasers

Laser stands for Light Amplification by Stimulated Emission of Radiation. The understanding of this process depends on understanding how light can interact with matter. We consider two energy levels i and j of a system which may consist of atoms, ions, or molecules. There are then three ways in which a photon of frequency ν, which has just the same energy as the gap between i and j, can interact.

(a) *Absorption*

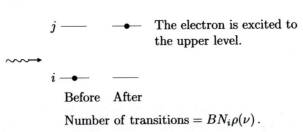

j ——   ——•—   The electron is excited to
                         the upper level.

i —•—   ——

Before   After

Number of transitions $= BN_i\rho(\nu)$.

The amount of absorption will depend on the number of particles in the lower state (N_i) and the density of the radiation, $\rho(\nu)$. The proportionality constant is the Einstein B coefficient.

(b) *Spontaneous Emission*

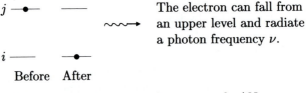

j —•—   ——   The electron can fall from
                        an upper level and radiate
                        a photon frequency ν.

i ——   ——•—

Before   After

The number of emitted photons equals AN_j.

There remains the third possibility that the incoming photon meets the particle in its excited state, i.e.

(c) *Stimulated Emission*

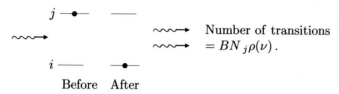

$$\text{Number of transitions} = BN_j\rho(\nu).$$

Before   After

In stimulated emission the incoming photon is augmented so that the two-outgoing photons are exactly in phase. The stimulated emission process is analogous to a phase transition and may be looked on as an example of the Bose-Einstein condensation encountered in the case of ^{4}He. The condensed photons obey Bose-Einstein statistics and consequently are able to congregate in a single energy level where they all have the same momentum. This gives the laser light its unique properties. To have stimulated emission as the major effect we must have N_j greater than N_i (a negative temperature) or the net effect will be absorption.

This population inversion is achieved in different ways in two of the currently widely used lasers. In the ruby laser, chromium atoms are 'pumped' by an intense pulse of light to one of the upper energy bands of the crystal. The atoms then give up some of their energy to the lattice and fall to a metastable level. It is the subsequent transition between this metastable level and the ground state which is involved in the laser action. The energy scheme for the ruby laser is shown in Fig. 8.5.

Alternatively, energy may be fed constantly (instead of in a pulse) into a system in such a way that two levels i and j which have allowed transitions between them always have $N_j > N_i$. The helium-neon laser is an example of this continuous working system. Helium is excited electrically and populates level j of neon by collision. State j cannot decay to the ground state as this is forbidden, but can decay by stimulated and spontaneous emission to levels i and i' which are initially unoccupied and which are easily depopulated by collision processes; an ideal solution for the stimulated process to predominate. The energy-level diagram is shown in Fig. 8.6.

Having obtained stimulated emission the second problem is to amplify it. This is achieved by having a cavity resonator. A wave which starts at one wall is reflected back and forth, being built up all the time. Any photon

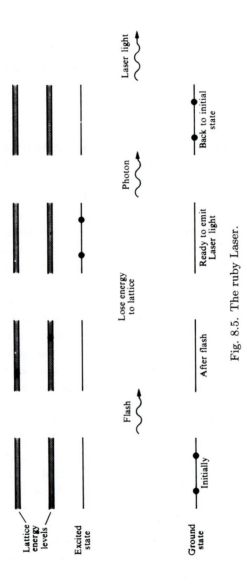

Fig. 8.5. The ruby Laser.

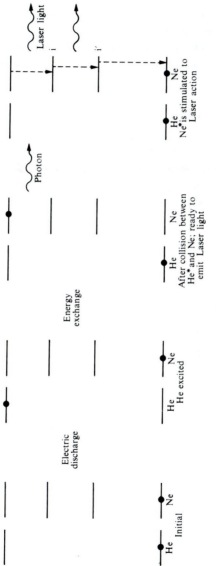

Fig. 8.6. The helium-neon Laser.

which travels in a direction not perpendicular to one of the mirrors leaves the system and is lost so that the light remaining in the resonator is all parallel and in phase. In practice confocal concave mirrors are usually used and one of them is made semi-transparent so that a small fraction of the light beam issues from the cavity.

As a consequence of this production by the stimulation process in a resonant cavity the output beam has these important properties, in addition to its high power.

It is (1) directional

(2) monochromatic

(3) coherent.

These properties give rise to the possibility of countless interesting experiments and applications which are of major current interest, yet the very heart of the laser is merely a distortion of the distribution of particles among their energy levels.

Chapter 9

COMPUTER SIMULATIONS

In principle, the properties of bulk quantities of molecules, as gases, liquids or solids, should be calculable from the properties of the individual molecules. If we know the forces between the molecules making up a substance then the properties of the complete ensemble should be computable. This is very much the point of view of statistical thermodynamics. The only snag, until relatively recently, was the fact that such a calculation would be wholly impracticable if the system contains enough particles to be a satisfactory representation of reality. If every particle interacts with every other one, then millions of interaction energies have to be calculated for a given arrangement of the particles.

Two advances have opened up this subject to the extent that it is now a thriving area of research with applications throughout chemistry and even molecular biology. The first obvious innovation has been the development of ever more powerful and less expensive computers capable of dealing with the vast number of computations required. The second has been the incremental improvements in empirical force fields: that is the so-called "molecular mechanics potentials" which are simple empirical formulae that give the energy of a molecule, or set of molecules, as a function of the distances between the constituent atoms.

9.1. Molecular Mechanics Potentials

Molecular mechanics potentials express intramolecular and intermolecular energy with mathematical functions in which the parameters are fixed by reference to experimentally observed properties from spectroscopy or by using purely theoretical quantum chemical calculations.

The intramolecular energy of a molecule will be a sum of terms representing bond stretching; angle variation; dihedral angles; non-bonded interactions and charge effects. These are summarized schematically in Fig. 9.1.

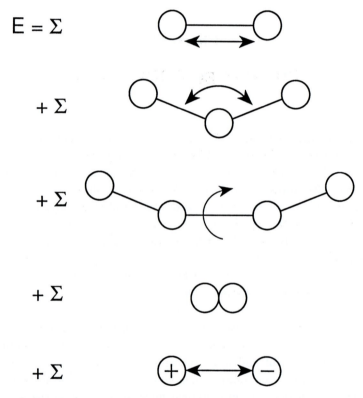

Fig. 9.1. Pictorial representation of the terms included in a molecular mechanics potential, Σ indicates the summation of terms due to stretching, binding etc.

Most force fields adopt a simple harmonic approximation for the bond stretch energy, with the energy varying as a bond length, l, changes from its equilibrium value l_o as the square of the stretch

$$E_l = k_1(l - l_o)^2 .$$

For greater accuracy a cubic term can be added. Similarly the energy variation with bond angle, Θ, is assumed to be harmonic

$$E_\Theta = k_\Theta (\Theta - \Theta_o)^2 \,.$$

In order to include dihedral angle (ω) variation, such as that which leads to a difference in energy between syn and anti ethane (Fig. 9.2), the functional form of a Fourier series is used

$$E_\omega = V_n (1 + s \cos n\omega)$$

with V_n being the torsional barrier, n the periodicity (e.g. $n = 3$ for ethane) and $s = 1$ for staggered minima or $s = -1$ for eclipsed.

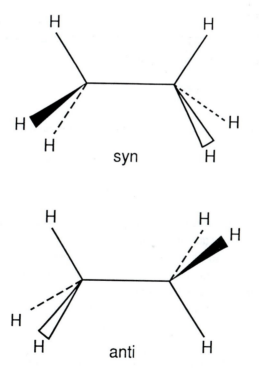

Fig. 9.2. Molecular geometries for syn and anti ethane.

Non-bonded interactions are often treated as if they are van der Waals terms

$$E_{\mathrm{vdw}} = \varepsilon \left\{ (r_m/r)^{-12} - 2(r_m/r)^{-6} \right\}$$

with ε the well depth and r_m the minimum energy interaction distance (Fig. 9.3).

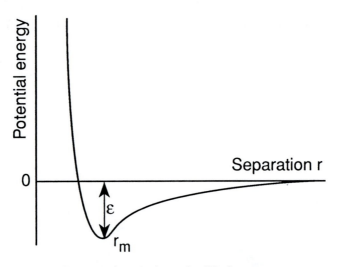

Fig. 9.3. A typical van der Waals curve.

The second non-bonded interaction component in the electrostatic term E_{el}

$$E_{\mathrm{el}} = q_i q_j / D r_{ij} \,.$$

The charges on atoms i and j, q_i and q_j, come from quantum mechanical calculations while the rather problematic dielectric constant, D, is sometimes given a bulk value and at others treated as having a distance dependence.

Further terms can be added to incorporate hydrogen bonding or inversion barriers, but in the end the entire energy expression contains perhaps several dozen parameters which have to be fitted so as to reproduce experimental energy results such as heats of formation or vibration frequencies.

Despite the crudity of this notion the potentials have been refined to the point where they are good enough to use in minimization routines that can predict the structures of proteins, given a rough starting structure, to a level which is perfectly acceptable, or to perform conformational analysis of a single molecule. However, if we are to treat large statistical assemblies of molecules, we need to have many molecules in the system in order to avoid surface effects. This is achieved using a simple trick called 'periodic boundaries'.

9.2. Periodic Boundary Conditions

Properties of a bulk phase of a pure compound, or the more commonly studied solute in a solvent, depend on having many thousands of molecules in our system. For example, experimental studies of molecules in water show that the water structure is influenced for several Ångstroms away from the solute (1 Å is 10 nm). At the same time there is a limit to just how many molecules even the fastest of computers can handle.

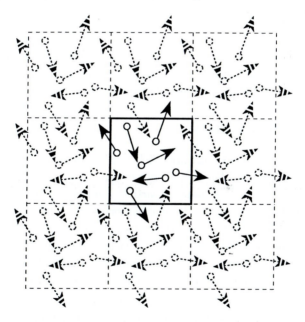

Fig. 9.4. Periodic boundary conditions. Solid lines represent the actual system studied and dotted lines the periodic images of that system.

If we carry out our calculations for a box with finite size, containing perhaps hundreds of molecules, then the atoms at the centre of the box will be experiencing their correct complement of interactions while those at the faces have a vacuum on one side. However, if we impose periodic boundary conditions, we surround our system with images of itself, as shown in Fig. 9.4.

The system becomes a type of unit cell in an infinite solution or crystal. Molecules at the edge of the real box can now interact with image molecules in the surrounding boxes, minimizing edge effects. If the motion of a molecule causes it to leave the simulation box, then an image enters from the opposite face, keeping the number of molecules in the system constant.

Care must be taken in the choice of box size and the cut-off distance of intermolecular potentials. The box length must be more than twice the cut-off distance to prevent a molecule A interacting both with B and the image B′ in an adjacent box. Since coulombic energies are quite long-ranging, varying as they do according to r^{-1}, these interactions have to be truncated.

Despite these difficulties, simulation with the periodic box of particles dense enough to mimic reality gives exciting results using one of two generally applicable approaches; Monte Carlo methods, or molecular dynamics.

9.3. Monte Carlo Methods

As the name suggests, this approach contains a random element as in gambling. We can calculate the energy of our set of molecules randomly placed in the periodic box; move a particle randomly and calculate again and again and again to find some average property. The snag about that simple idea is that all sets of positions of the particles would be treated as equally probable, but we know from the Boltzmann distribution law that some configurations of the molecules are more likely than others. This use of the Boltzmann distribution idea is the link with statistical thermodynamics.

In practice, we may, for example, be dealing with a solute in a periodic box of discrete solvent molecules, randomly placed but with a density which is a true reflection of solvent density. The energy of the system is computed using our molecular mechanics potential. One molecule is picked at random, moved randomly in three dimensions and randomly rotated. The new energy of the system is evaluated. If this is lower than the previous position, that new configuration is added to the set over which an average will be taken for the property of interest. If the newly generated configuration has a higher energy, its Boltzmann probability is evaluated,

$$\exp(-\Delta E/kT)\,.$$

This is compared to a randomly generated number between 0 and 1. If the probability is higher than the random number the new configuration is accepted, otherwise it is rejected and the system is returned to its previous state. By repeated, perhaps by millions of applications of this procedure, a correctly Boltzmann weighted set of configurations is obtained (including some which are quite unlikely or rarely occurring). If we have M such configurations the average value of some property, Q_{av}, is simply given

$$Q_{av} = \frac{1}{M} \sum Q(x')$$

where x' indicates only configurations with an acceptable Boltzmann weighted probability. This is sometimes referred to as Metropolis sampling.

Our initial guessed arrangement of particles will probably have a high energy but this will reduce as the number of moves increases during these early equilibrations until it reaches some sort of consistency. These early configurations are not included in those which go to make up the 500,000 to one million subsequent examples over which averages are taken.

Monte Carlo calculations are very good at searching over a complicated potential surface, moving from one minimum to another. They are less suitable for dealing with macromolecules such as proteins since most moves of atom position generate very unlikely arrangements. Neither is the technique suitable for the study of time-dependent phenomena such as diffusion. Time dependent properties of systems can, on the other hand, be studied using molecular dynamics.

9.4. Molecular Dynamics

In molecular dynamics we try to simulate the motions of a system of atoms with respect to the forces which are present. Each configuration follows from the previous one as a result of internal forces, not as in Monte Carlo simulations by external generation of random changes. In effect what is done in molecular dynamics is to solve Newton's equations of motion for the particles just as one might for a set of billiard balls, but involving many thousands of atoms. It is the collective forces which cause the system to change with respect to time.

From Newton's second law, the force F_i on particle i of mass m_i is given by

$$F_i = m_i a_i$$

with a_i being its acceleration. It we write the acceleration, a_i, as the second derivative of displacement (x) with respect to time,

$$a_i = \frac{d^2 x_i}{dt^2}$$

so

$$\frac{d^2 x_i}{dt^2} = \frac{F_i}{m_i}.$$

To follow the dynamic behaviour of our system, we must solve this equation for every particle in the system. On integration with respect to time, we get

$$\frac{dx_i}{dt} = (F_i/m_i)t + c_1.$$

At time $t = 0$ the first term vanishes and the velocity is given by the constant c_1, the initial velocity u_i. At time t, we have

$$\frac{dx}{dt} = a_i t + u_i$$

the expression for the velocity at any time. On further integration with respect to time

$$x_i = u_i t + a_i t^2 / 2 + c_2$$

where the constant c_2 is the current position. This allows one to calculate the displacement from the initial velocity u_i and the acceleration $a_i = m^{-1} F_i$; the force coming from the molecular mechanics potential.

This simple derivation gives us an expression which corresponds to a Taylor series for the displacement

$$x_{(t+\Delta t)} = x_t + (dx/dt)\Delta t + (d^2 x/dt^2)\Delta t^2 / 2 + \ldots$$

and we neglect higher terms. We also assume that the acceleration remains constant over the timestep, Δt. In practice, timesteps are usually taken as 0.5–1 Femtoseconds (1 Fs $= 10^{-15}$ sec). To minimize errors due to using a finite timestep and truncating the Taylor series, devices such as the "Verlet leapfrog" method are employed. In this method we write v for the average velocity over a timestep Δt. Then the displacement after the interval Δt is given by

$$x_{(t+\Delta t)} = x_t + v\Delta t$$

omitting the higher terms in the Taylor series because Δt is so small. Assuming v is almost equal to the velocity at the midpoint of the time interval,

$$v = dx/dt \quad \text{at} \quad (t + \Delta t/2).$$

We can now calculate v from the midpoint of the previous interval and average the acceleration between $t - \Delta t$ and $t + \Delta t$, so

$$v_{(t+\Delta t/2)} = v_{(t-\Delta t/2)} + a\Delta t$$

where a can be calculated from $m^{-1}F(x,t)$. Hence the new position is given by

$$x_{(t+\Delta t)} = x_t + v_{(t+\Delta t/2)}\Delta t.$$

The actual systems studied in this way are essentially the same as those in Monte Carlo calculations. The system is minimised and then heated to the required temperature. The temperature of the simulation is calculated from the kinetic energy of all the atoms of the system,

$$1/2 \sum_i m_i v_i^2 = 3/2 \, NkT.$$

At time $t = 0$ no velocities are known so the system is heated by assigning velocities randomly according to a Maxwellian distribution for the appropriate temperature. Usually a period of some picoseconds of simulation is allowed for the system to equilibrate as in Monte Carlo work and then the production dynamics is carried out, over which time averages may be calculated or behaviour followed with respect to time. Current computer power normally limits simulation to time periods of at most one nanosecond, but this is quite a long time on a molecular time scale.

9.5. Free Energy Calculations

In much of chemistry and even more in biochemistry, the key measurements are of free energy changes, derived from equilibrium constants. Using the results from Chapters 1 and 3, we can see that free energies are simply related to partition functions. In the simplest case of a system of localized particles, $S = k \ln q^N$; $Q = q^N$ and $A = U - TS$.

$$\therefore A = -kT \ln Q.$$

If we know the energy levels of a set of non-interacting and indistinguishable particles, then as we saw in Chapter 4 the partition function can be computed by summing over the energy levels

$$Q = q^N/N! \quad \text{and} \quad q = \sum_i p_i \exp[-e_i/kT].$$

However for systems of many particles which do interact in ways given by our molecular mechanics potentials, it is not feasible to determine the quantum mechanical energy levels and it is necessary to treat the system classically.

The classical energy is given by the classical Hamiltonian H, the sum of kinetic and potential energies and is a function of the position, r, and momentum, p, of every particle. $H(r,p)$ is thus a continuous function and the summation in the expression for Q becomes an integral

$$Q = \frac{1}{N!}\frac{1}{h^{3N}} \int \cdots \int \exp[-H(r,p)/kT] \, dr \, dp.$$

The constant $1/h^{3N}$ in front of the integral is the unit of 'phase space', $r \times p$ (position $\times$ momentum) in three dimensions. We know from the uncertainty principle (position $\times$ momentum) has the units of h, (action) and here we have N particles in three dimensions. Even now we cannot normally evaluate Q and hence free energy, but differences are obtainable. For an equilibrium, a chemical equilibrium for example

$$X \rightleftharpoons Y; \qquad K = [Y]/[X]$$
$$\Delta A = -kT \ln(Q_Y/Q_X)$$
$$= -kT \ln \left[\frac{\int \cdots \int \exp[-H_Y/kT] \, dr \, dp}{\int \cdots \int \exp[-H_X/kT] \, dr \, dp} \right].$$

Now if we assume that in going from X to Y only a small change or perturbation has been made, such that

$$H_Y = H_X + \Delta H$$

then

$$\Delta A = -kT \ln \left[\frac{\displaystyle\int \cdots \int \boxed{\exp[-H_X/kT]} \exp[-\Delta H/kT]\, dr\, dp}{\displaystyle\int \cdots \int \exp[-H_X/kT]\, dr\, dp} \right].$$

This seemingly complicated expression becomes both simple and useful if we note that the portion in the box in dashed outline is just the Boltzmann distribution law, or a probability. Hence

$$\Delta A = -kT \ln \langle \exp[-\Delta H/kT] \rangle_X$$

with the $\langle \cdots \rangle_X$ indicating the average value of the perturbation taken over configurations generated for the system X. These may be derived by either Monte Carlo or molecular dynamics runs. Once the many configurations have been generated then free energy difference can be evaluated by averaging the $\exp[-\Delta H/kT]$ terms calculated for each configuration. If the simulations are done under constant pressure conditions (instead of the constant volume conditions applying to ΔA) then we can derive Gibbs free energy differences, ΔG; the most useful of all thermodynamic quantities.

In this way, the so-called free energy perturbation technique, it has proved possible to obtain differences in free energies of solution between different compounds; differences in partition coefficients; redox potentials and perhaps most excitingly the relative binding free energies of different inhibitors to an enzyme or different drugs to the macromolecular receptor to which they attach themselves.

There is no doubt that the availability of computers has made statistical thermodynamics one of the most vigorous and promising of current research areas.

Chapter 10

SOLIDS AND BIOPOLYMERS

The imaginative application of the ideas and methods of one area of scientific thinking to some other branches of endeavour can be a stimulating exercise. In this final chapter we shall be looking at some of the ways in which our earlier statistical notions of the interplay between energy levels and the distributions of particles can be extended, not necessarily rigorously, to a diversity of scientific and non-scientific situations. This is not intended to be a comprehensive selection and there are other, no less interesting situations (for example, the coiling of polymer chains in solution) which are susceptible to treatment on similar lines. We shall be particularly concerned with solids in which there may be an entropy associated with the random orientations of the constituent molecules or groups with respect to one another and with biopolymers which can be considered as one-dimensional solids. The third law then predicts that, provided internal thermodynamic equilibrium is maintained, configurational entropy will disappear at low enough temperatures. As usual, thermodynamic considerations give us no guide to the nature of the ordering nor to the mechanism by which the ordering process will take place. It turns out in practice that internal thermodynamic equilibrium is by no means always maintained as a system is cooled. When this happens, one or more aspects of the system may retain some entropy even at the lowest temperatures. Instead of the excellent agreement which is usually found between calculated and observed entropies, illustrated in

Chapter 6, there is then a discrepancy; the calculated entropy is greater than the entropy obtained from heat capacity data.

Now let us examine the general considerations which will determine whether an aspect for a system is likely to be able to maintain itself at equilibrium as the temperature is lowered. We will take as a schematic model a linear molecule which can have one of two positions along a line. The ordered state is represented by molecules which are oriented head-to-tail and the disordered state by a random arrangement of molecules.

From the third law we know that the ordered state is thermodynamically favoured at low temperatures.

The orientations shown as ← and → are the only two stable lattice configurations for the molecule; any intermediate position such as ↑, in which the axis of the molecule is inclined at an angle to the row by hypothesis is

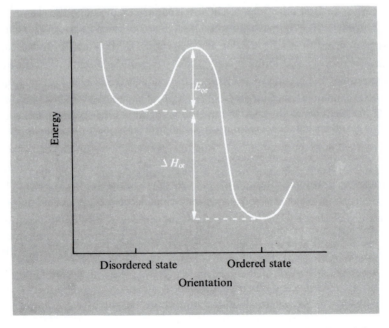

Fig. 10.1. Potential-energy diagram of motion in a molecular solid.

of higher energy. Thus, in turning over from one direction along the axis to the other direction the molecule must go over an energy barrier. If we now call the magnitude of this barrier to orientation E_{or} and the energy difference between the ordered and the disordered state ΔH_{or}, we can draw the schematic potential-energy diagram shown in Fig. 10.1.

We are interested in two parameters of the system. First, the distribution of the molecules between the ordered and the disordered states and second the rate at which the molecules are able to cross the energy barrier E_{or}, between the two states. At high temperatures the molecules will be passing back and forth across the barrier readily and equilibrium will be maintained in accordance with the distribution law. In simplified form:

$$n_{\text{disordered}} \propto \exp\left(-\Delta H_{or}/RT\right). \qquad (10.1)$$

The rate at which the molecules pass from left to right (i.e. disordered to ordered) is described by the Arrhenius equation, which gives the rate constant for the process as:

$$k = A \exp\left(-E_{or}/RT\right). \qquad (10.2)$$

When the system is cooled the molecules will have a tendency to congregate in the ordered state in accordance with Eq. (10.1). At the same time their rate of passage across the barrier from the disordered to the ordered state diminishes in accord with Eq. (10.2). Thus, both the parameters in which we are interested have an exponential dependence on temperature. Depending on the circumstances one or other of these exponential factors may dominate the situation. We can identify two particularly interesting and important circumstances, as follows:

(i) $E_{or} \ll \Delta H_{or}$. In this case, cooling the system leads to an increase of population in the ordered state of temperatures where the rate of crossing the barrier is still high. It is then kinetically possible for equilibrium to be maintained throughout the cooling process and the system will be well-behaved i.e. $S \to 0$ as $T \to 0$.

(ii) $E_{or} \gg \Delta H_{or}$. The barrier height is now much greater than the thermodynamic difference between the two states. This high barrier will inhibit transfer from the disordered to the ordered state at comparatively high temperatures. However, at these temperatures the distribution law, Eq. (10.1) predicts that there will still be many molecules in the disordered state. By the time that further cooling has caused the distribution to shift in favour of a predominantly ordered state the rate of interchange has been reduced to a negligible level. The molecules have thus effectively become trapped in their disordered configuration. Over the time scale of an experiment the entropy associated with the disarray is then retained by this aspect down to the lowest temperature. There will then be a difference between the calculated and observed entropies for this system with $S > 0$ as $T \rightarrow 0$.

We thus come to the conclusion that when there is a discrepancy between calculated and experimental entropies this need not be attributed to a failure or even to an apparent failure of the third law. Rather, it indicates that the central requirement, for internal thermodynamic equilibrium, has not been fulfilled. It is true that if it were possible to wait a sufficiently long time the molecules might be able to align themselves and thus to eliminate the residual entropy. However, the experimentalist has only a limited time available and on this conventional time scale the events will be as described. We may note in passing that there are plenty or familiar examples of non-equilibrium situations, even on the celestial time-scale; for example, the rate of combustion of many fuels at room temperature is immeasurably low.

Having identified the general features of a molecular solid that are likely to lead to a residual entropy at the lowest temperatures, we are now in a position to look for some examples. As a start we look for molecules which are similar to our linear model and in which the intermolecular forces are comparatively insensitive to orientation. There will then be, as we require, a relatively small energy difference between the ordered and disordered states i.e. ΔH_{or} is small. Two simple molecules that fulfill these conditions are carbon monoxide and dinitrogen oxide. Both molecules have small dipole moments ($\mu_{co} = 0.11$ debye $= 3.6 \times 10^{-31}$ Cm; $\mu_{N_2O} = 0.17$ debye $= 5.6 \times 10^{-31}$ Cm) with the result that dipole-dipole forces within the lattice are weak compared with the other intermolecular forces present. These latter forces do not change their sign when the orientation of one molecule relative to another is changed by $180°$. The lattice energy is, therefore, insensitive to the orientation of the molecules along the axis. In accordance with these expectations both carbon monoxide and dinitrogen oxide probably have residual entropies at the absolute zero.

It is possible to make an estimate of the entropy associated with the orientational process, the entropy which may be frozen in at low temperatures. For the two-position model used so far there are 2^N ways of arranging the N molecules of one mole, i.e.

$$W = 2^N .$$

Therefore

$$S_{\text{residual}} = k \ln W = k \ln 2^N = R \ln 2$$
$$= 5.8 \text{ J K}^{-1} \text{ mol}^{-1} .$$

This is the maximum of the residual entropy, a value which will be observed only when no ordering at all occurs during cooling. For both carbon monoxide and dinitrogen oxide the residual entropy is 4.9 J K^{-1} mol^{-1}, a result that is close enough to 5.8 J K^{-1} mol^{-1} to support the proposed explanation, but suggests that some ordering has taken place.

The residual entropy in our simple linear example is, as we have shown,

$$S_{\text{residual}} = k \ln 2^N ,$$

there being N particles and one of two possible orientations. Another way of looking at this result is to consider it as a measure of the lack of *information* one has about the orientations of the molecules. This notion of relating entropy to a lack of information will be considered further later in this chapter when we deal with biopolymers.

We can extend our simple model to include any structure in which entropy is associated with a re-orientation process within the lattice. A long-standing example is the residual entropy of ice. Instead of the more usual close-packed array of molecules found in molecular lattices the basic structure of ice is much more open and has the oxygen atoms in a tetrahedral configuration. Each oxygen atom is chemically bonded to two hydrogen atoms at a distance of 0.10 nm and hydrogen-bonded to two hydrogen atoms, at a distance of 0.18 nm. One hydrogen atom lies between each pair of oxygen atoms, and is chemically bonded to one oxygen and hydrogen-bonded to the other. This open structure reflects the influence of hydrogen-bonding on molecular lattices and it causes ice to have a low density. Now there are many ways in which the hydrogen atoms can distribute themselves in the lattice while fulfilling the requirement of two bonded and two hydrogen-bonded atoms per oxygen atom. It has been shown that the configurational entropy associated with the hydrogen atoms is $k \ln(\frac{3}{2})^N = R \ln \frac{3}{2} = 3.39$ J K^{-1} mol^{-1}. This is close to the experimental value for the residual entropy, 3.36 J K^{-1} mol^{-1}. A more recent discussion has been concerned with the residual entropy of 'sulphuryl fluoride' SO_2F_2. It is not unexpected that this molecule should have a residual entropy, since it has a small dipole moment, 0.23 D ($= 7.6 \times 10^{-31}$ cm) and is nearly tetrahedral. For so large a molecule the accurate calculation of the statistical entropy is difficult, but the discrepancy between the calculated value and the third law experimental value is close to $R \ln 2$. A possible explanation here is that the dipoles do line up axially but that there are two molecular orientations along the axis which are taken up at random.

λ-Points and Second-Order Transitions

In the discussion so far we have used as our model a system which transforms over a range of temperature from the disordered to the ordered state. However, many transformations in the solid state occur at a definite temperature. Indeed when the transformation is between two identifiable separate phases it is required by the Phase Rule to take place at a particular temperature. Since

$$P + F = C + 2$$

and for a single compound $C = 1$, the number of phases $P = 2$ so that the number of degrees of freedom, $F = 1$. This is used up in defining the external prressure, with the result that the temperature must be constant. When the difference between the states is less pronounced, the transformation may occur less sharply but still over a considerably narrower temperature range than is involved in our earlier discussion. We can see how this situation might arise if we postulate that the environment and, therefore, the energy of a molecule depends significantly upon the orientations of the adjacent molecules. At low temperatures the molecules are lined up in an ordered array in accordance with the third law. As the substance is warmed up increasing thermal motion may cause one molecule to move out of alignment, producing a region in which the regularity of the field due to intermolecular forces has been reduced. The result of this may be to make it easier for an adjacent molecule also to become disoriented, and so on through the lattice. There is, thus, a co-operative spread of disorder through the lattice, which may happen over a comparatively narrow range of temperature.

The energy required for this randomization process will be taken in over the same temperature range. Then because $C_V = (\partial U / \partial T)_V$ the heat capacity will temporarily become much greater than its normal value. Once disorder is complete no further energy is required by this aspect of the system and the heat capacity falls back to a normal value. The graph of heat capacity versus temperature has the form shown in Fig. 10.2. The resemblance of this curve to the Greek letter lambda (λ) has caused the transition to be known as a *lambda point*. An alternative way of looking at gradual transitions has been to contrast them with the abrupt transitions that occur at phase changes and to call the latter 'first-order transitions' and the former 'second-order transitions'.

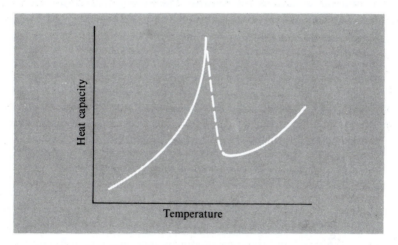

Fig. 10.2. Heat capacity at a lambda-point transition.

Second-order transitions are few enough in number that it is probably better to consider each one separately and to try to deduce the individual molecular motions that give rise to the λ-point rather than to aim at producing a comprehensive general theory. We shall now discuss some examples of λ-point transitions.

Ammonium Chloride

One λ-point transition which has attracted much interest occurs in solid ammonium chloride at 242.7 K. The crystal structure of this salt near the λ-point is body-centred cubic with chloride ions at the corners of the cube

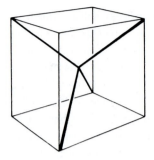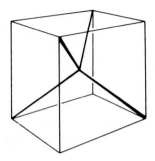

Fig. 10.3. The two orientations of the NH_4^+ ion in NH_4Cl solid.

and the ammonium ions at the centres. The hydrogen atoms of the tetrahedral ammonium ions point toward the Cl^- ions. There are, however, two
energetically equivalent positions for the NH_4^+ tetrahedron in this structure.
These are illustrated in Fig. 10.3.

There can thus be a configurational entropy associated with the two
possible orientations of the NH_4^+ ions. At low temperatures this configurational entropy is lost because all the ammonium-ion tetrahedra line up with
their axes parallel. It is worth noting that this is by no means the only way
in which the configurational entropy can be lost. In ammonium bromide,
it is thought, the NH_4^+ ions are aligned at low temperatures with alternate ions in opposed orientations. When the ammonium chloride crystal
is warmed up the oriented ammonium ions execute vibrational oscillations
about their mean positions. There comes a point at which some ions flip
over into the other configuration. The field experienced by their neigh-
bours is then altered and these neighbouring ions can in turn flip over more
readily. Orientational disorder thus spreads throughout the crystal over
a narrow temperature range. Energy is required to produce this disorder
and a λ-point is observed. Once the orientations of the tetrahedra have
been randomized the heat capacity returns to a normal value. It should be
noted that above the λ-point the evidence from heat-capacity data shows
that the NH_4^+ ions are *not* freely rotating, but are still executing vibration
oscillations, about random orientations.

In addition to the thermodynamic data on which these conclusions are
based there is confirmatory evidence from nuclear magnetic resonance experiments. In the solid state, n.m.r. line-widths and relaxation times are
both sensitive to molecular motion. Thus a change from a vibrational

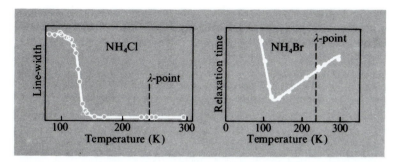

Fig. 10.4. N.m.r. properties at λ-points of NH_4Cl and NH_4Br (in solids the line width is often described by the 'second moment' which is approximately equal to the square of the line width at half the peak height).

motion to free rotation of the ammonium ions would produce a marked change in both n.m.r. parameters. Experimentally no discontinuity in either property is found on passing through the λ-point as Fig. 10.4 shows for ammonium chloride and bromide.

Some Antiferromagnetic Compounds

There is frequently a λ-point associated with the magnetic transition from the paramagnetic to the antiferromagnetic state. Antiferromagnetism, as we saw in Chapter 7, is due to the ordering of the magnetic moments associated with the unpaired electrons present in the ions of compounds of transition and rare-earth metals, mostly oxides and halides. The ordering takes place by a pairing process and thus tends to make the compound diamagnetic. The change from paramagnetism to antiferromagnetism can be depicted schematically

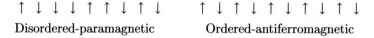

Disordered-paramagnetic   Ordered-antiferromagnetic

Paramagnetic compounds follow the *Cuire-Weiss Law*, so that the dependence of their magnetic susceptibility is of the form

$$\chi = \frac{C}{T - \Delta}$$

Where C and Δ are constants for a particular compound. However, below some critical temperature, the Néel point (see p. 94), the susceptibility may

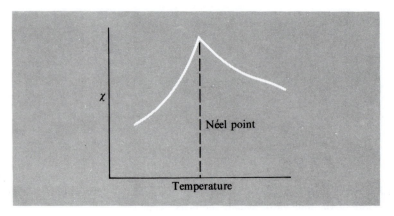

Fig. 10.5. Temperature-dependence of the magnetic susceptibility of an antiferromagnetic compound.

fall dramatically as a result of the ordering process. The susceptibility may, therefore, pass through a sharp maximum at the Néel temperature, as shown in Fig. 10.5.

If now we consider what happens as the compound is warmed up from a low temperature, we find that the heat capacity is normal in the antiferromagnetic region. At the Néel point there is a rapid increase in magnetic disorder over a narrow temperature range. This process requires internal energy with a resultant rise in the heat capacity. Once the random, paramagnetic state is reached, no further internal energy is required for this process and the heat capacity returns to a value near its former level.

Biopolymers

The long chains of organic molecules which make up biopolymers make it reasonable to consider, as a first approximation, these polymers as one dimensional solids; the basic units repeat in a linear sequence rather than in three dimensions. Of particular interest are the nucleic acids (DNA and RNA) and proteins. The former are made up of linearly joined nucleotides (Fig. 10.6(a)) and the latter from amino acids (Fig. 10.6(b)).

The work of molecular biologists has shown that it is the DNA in cells which contains the genetic information required to build protein polymers. One is then naturally led to question whether this information content of molecules can be treated by the methods of statistical thermodynamics.

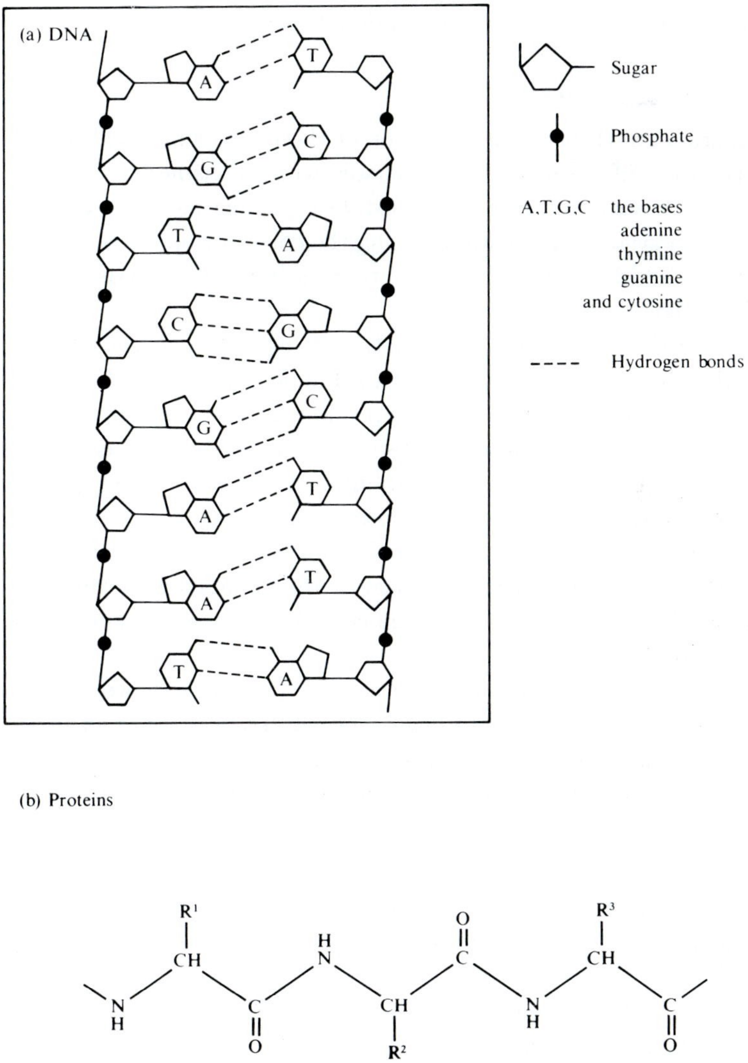

R[1], R[2], R[3] etc represent amino acid residues

Fig. 10.6. Schematic representation of (a) DNA (b) Protein.

By an extension of our statistical ideals it can indeed be discussed, using a branch of the subject called Information Theory.

Information Theory

Information theory is an extension of thermodynamics and probability theory. Much of the subject is associated with the names of Brillouin and Shannon. It was originally concerned with passing messages on telecommunication systems and with assessing the efficiency of codes. Today it is applied to a wide range of problems, ranging from the analysis of language to the design of computers.

In this theory the word 'information' is used in a special sense. Suppose that we are initially faced with a problem about which we have no 'information' and that there are P_0 possible answers. When we are given some 'information' this has the effect of reducing the number of possible answers and if we are given enough 'information' we may get to a unique answer. The effect of increased information is thus to reduce the uncertainty about a situation. In a sense, therefore, information is the antithesis of entropy, since entropy is a measure of the randomness or disorder of a system. This contrast led to the coining of the word *negentropy* to describe information.

The basic unit of information theory is the *bit*—a shortened form of 'binary digit'. Suppose we start with no information about an initial situation of which there are P_0 equally possible outcomes, and we are then given enough information to specify which outcome is realized, i.e. $P_1 = 1$. Then:

$$\text{Initially}: \quad I_0 = 0, \quad P_0 \text{ outcomes}$$
$$\text{Finally}: \quad I_1 \neq 0 \quad P_1 = 1.$$

For example, if one is given a playing card face down without any information, it could be any one of 52; if one is then told that it is an ace, it could be any one of 4; if told that it is also a spade, one knows for certain which card one has. As we are given more information, The situation becomes more certain. In general, to determine which of the P_0 possible outcomes is realized, the required information I_1 is defined as

$$I_1 = K \ln P_0.$$

If we treat our method of narrowing the choice as a problem of making n independent binary selections (such as in a linear crystal where each molecule could line up in one of two directions) then

$$P_0 = 2^n$$

Therefore

$$I = K \ln P_0 = Kn \ln 2 \,.$$

Now we want to identify the information I with the number of binary digits n, so we take $K = 1/\ln 2 = \log_2 e$

$$I = \log_2 P$$

or if we have P_0 initial possibilities and infer information I the possibilities are reduced to P, then the amount of information we have been given is

$$I = \log_2 \frac{P}{P_0} \,.$$

A simple example may suffice to illustrate this. If we have to identify one specific number 1 and 32, the information required to do so is

$$I = \log_2 32 = 5 \text{ bits of information} \,.$$

One could consider these five binary choices in the following way. Suppose we were asked to find a specified number between 1 and 32 and could ask questions to which we could only be told the answer yes or no. The information is then being given to us in binary form. We might pose questions as follows:

'Is the number in the first half of the list 1–32?'

Given the answer yes or no, we could then say:

'Taking that half of the numbers is the number in the first half of these?' Thus we would continue, but after five answers, i.e. five bits of information, we could be guaranteed to have the correct number.

We have above defined the information as $I = K \ln P$. The similarity of this equation to our definition of entropy (in Chapter 2) as $S = k \ln W$ arises because both entropy and information (or negentropy) are additive. We can make the link closer if we choose to make the constant K the Boltzmann constant k; we will then obtain I in entropy units, a point to which we will return again.

Information in Biopolymers

First let us consider the information content of biopolymers. The molecular weight of a typical protein is about 120,000. The molecule will contain something like 1000 amino acid residues each being one of the 16 naturally occurring acids. Following what we said above, we will need 4 bits (since $16 = 2^4$) of information to specify each acid.

$$\text{Therefore } I_{\text{protein}} \approx 4 \times 10^3 = 4000 \text{ bits}.$$

Now let us consider a typical DNA molecule. This would normally have a molecular weight of about 10^6 and be build from 4000 nucleotides each being one of the four basic units, adenine, thymine, guanine or cytosine, i.e.

$$I_{\text{DNA}} = 4000 \times 2 = 8000 \text{ bits}.$$

Hence at this very simple level of understanding, a protein contains more information for a given molecular weight than does the DNA which codes it. This is consistent with the findings of Crick and Watson that the code is a triplet code with combinations of three nucleotides being required to code for each amino acid in the protein.

Crude calculations suggest that a single cell contains something like 10^{12} bits of information, estimated in the way sketched out above. This may be compared with a similar crude calculation which indicates that the Encyclopaedia Britannica contains approximately 10^{10} bits, an observation which does not belittle the encyclopaedia, but rather makes one realize why the cell is such complex structure.

What have these notions of information got to do with entropy? The connection is made for us by an engaging fellow called Maxwell's Demon.

Maxwell's Demon

Maxwell conceived of the following hypothetical experiment. Suppose we have two vessels (A and B) connected by a frictionless trap door, and that we start with a gas in one of the vessels. The gas can be thought of as consisting of hot (O) and cold (X) molecules, judged by their speeds. See Fig. 10.7.

Maxwell's Demon sits by the trap door and opens it to allow any fast (hot) gas molecules to pass. Since the door is frictionless he does no work

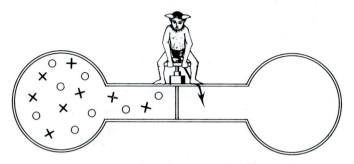

Fig. 10.7. Maxwell's demon.

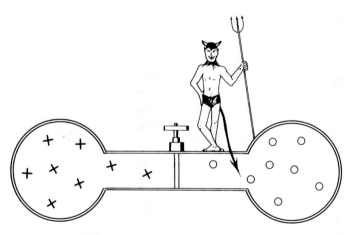

Fig. 10.8. Maxwell's demon: later.

in merely opening the door. After he has been doing this for a while the situation shown in Fig. 10.8 will be reached.

Even though he has done no work we have gone from a random situation to one which is considerably more ordered. This is quite contrary to the second law of thermodynamics whichever way one looks at it; for instance having achieved a hot body from a cooler one it would be a simple matter to produce a variety of perpetual motion devices. The resolution of this paradox, the exorcism of the demon, has been the occasion of much scientific ingenuity.

Szilard in the late 1920s pointed out that although the Demon does no work, he does require information. He postulated that the information once obtained is available to decrease the entropy of the system.

$$I \equiv -\Delta S, \qquad \text{or information} \equiv \text{negentropy}.$$

We can link the ideas of entropy and information by calculating how much information is required to define the orientations of the linear molecules in the ordered state we discussed earlier in this chapter. Because each molecule can have two orientations in the crystal one bit of information is required for each molecule. We can then write

$$P_0 = 2^N$$

and in the hypothetical, aligned state,

$$P_1 = 1.$$

Therefore

$$I = \log_2(P_0/P) = N \text{ bits}.$$

Thus in order to know that the ordered state has been achieved we require N bits of information for one gram molecule. If we wish to express this information in entropy units we compare the equations

$$S = k \ln W$$

$$I = K \ln P$$

and note that $K = 1/\ln 2$ and $k = 1.38 \times 10^{-23}$ J K^{-1}.

$$k/K = k \ln 2 \simeq 10^{-23} \text{ J K}^{-1}.$$

Thus to express I in entropy units we multiply by 10^{-23}. For our simple solid:

$$I \simeq 6 \times 10^{23} \times 10^{-23}$$

$$I \simeq 6 \text{ J K}^{-1}\text{mol}^{-1}.$$

This value of I is close to the observed residual entropy of dinitrogen oxide, as it should be. That is to say in the absence of thermodynamic equilibrium we have no information about the orientations of the molecules.

Limitations on Ideas of Information Content

These ideas of information content give us some insight into the way in which development is coded by DNA which directs the production of proteins. However, care should be taken to ensure that these ideas are not pushed beyond their areas of applicability. It seems likely that the genetic coded material goes far beyond giving a programme for the synthesis of proteins. It may give information about the time sequence of events and higher order structure and function. This being so we clearly cannot treat DNA as a sort of simple computer tape. One codeword may specify alanine, and we can give a number to the amount of information thus given (4 bits). Each amino acid however may have many biological roles depending on what part it plays in biological function.

The concepts and terminology of information theory should be kept for the areas where they are appropriate. Medawar in his essay on Herbert Spencer [see suggestions for further reading] comes out strongly against the use of information theory to explain biological orderliness. He correctly points out that some of the difficulties are due to the use of words like 'order' in several quite different contexts. Biological macromolecules have a variety of roles which are at present beyond quantifying. For example, as well as acting as an enzyme a protein molecule may have a structural role, it may fetch and carry substrates and at the same time act as an insulator or alter solubility. Information theory may shed some light on these diverse patterns, but it cannot explain everything in a quantitative fashion.

Linguistics

When we go beyond the level of cells and the machine-like qualities of living systems and consider the artistic activities of human endeavour we find that even here statistical ideas of thermodynamics have a role to play, even in non-random selection where human intelligence intervenes.

We can calculate linguistic entropy in speech, music, or writing in the manner which people approach code cracking. If we consider a series of letters A F G L D S J ..., it is possible to compute the chance of a particular series coming up and to express this as the information content of the message or its negentropy. This will differ from language to language and some letters crop up more frequently than others, for instance the letter E in English.

A more refined analysis works in terms of component syllables. One can test what is significant in a syllable in speech by swapping syllables and seeing if meaning or tense is changed or lost. The table gives some examples of the application of this statistical approach to some works of literature.

AUTHOR	WORK	NUMBER OF SYLLABLES PER WORD	ENTROPY
Shakespeare	Othello	1.29	0.29
Galsworthy	Forsyte Saga	1.34	0.33
Thomas Mann	Buddenbrooks	1.74	0.51
Sallust	Epistula II	2.48	0.64

The type of interesting results that arise from such studies include: (a) English has the lowest entropy of any major language, and (b) Shakespeare's work has the lowest entropy of any author studied.

These ideas are now progressing beyond the scientific level and are impinging on new ideas of criticism. Here as in biology, the thermodynamic notions can be helpful though they must be applied with caution because concepts such as 'quality' cannot be measured as they are purely subjective.

Conclusion

Thermodynamics started with engineering devices and has now progressed through statistical ideas to influence even literature. A salutory place to stop is to turn full circle and end with a beautiful sixteenth century French love poem by Maurice Scève, called *The water clock*. It describes the thermodynamic cycle of a water clock. The mechanism consists of a lover's tears falling on his burning breast, evaporating only to condense on his icy forehead to run down again and complete the cycle.

> L'humidité, Hydraule de mes yeulx
> Vuyde tousjours par l'impie en l'oblique
> L'y attrayant, pour air de vuydes lieux
> Ces miens souspirs, qu'à suyvre elle s'applique
> Ainsi tous temps descent, monte, et replique
> Pour abrever mes flammes appaisées

Doncques me sont mes larmes si aisées
A tant pleurer, que sans cesser distillent?
Las! du plas hault goutte a goutte elles filent
Tombant aux sains, dont elles sont puysées.[1]

MAURICE SCÉVE, c. 1500–1564

---

[1]For those readers who have difficulty with sixteenth century French, a fairly literal translation is as follows:

As a result of her indifference, the tears stream from my eyes as from a water clock, drawn into the cycle as air into empty space by these sighs of mine which they hasten to pursue. So they continually fall, rise, then fall again to quench and damp the ardour of my love. Is it so easy to weep that my tears fall steadily, one by one? Alas, drop after drop they trickle down from above and fall onto the breast where they evaporate and are once more drawn up.

APPENDIX 1: SUMMARY OF TRANSLATIONAL CONTRIBUTIONS TO THERMODYNAMIC FUNCTIONS

For convenience we summarize the results obtained in Chapter 4 in forms useful for calculation, and include the other common thermodynamic functions. In all cases the results are obtained using the formulae of that chapter and substituting the molar translational partition functions. They represent the complete contributions for monatomic gases or the translational part for polyatomic gases. (Subscript P and V denote constant pressure or volume conditions.)

$$U = 3/2\,RT \qquad H = 5/2\,RT$$

$$C_p = 5/2\,R \qquad C_V = 3/2\,R$$

$$S = R \ln\left(\frac{2\pi\,mkT}{h^2}\right)^{\frac{3}{2}} \frac{V}{N} e^{\frac{5}{2}} = R \ln\left(\frac{2\pi\,mkT}{h^2}\right)^{\frac{3}{2}} \frac{kT}{P} e^{\frac{5}{2}}$$

$$= R(3/2 \ln M + 3/2 \ln T + \ln V + K_{SV})$$

$$= R(3/2 \ln M + 5/2 \ln T - \ln P + K_{SP}).$$

In these last two equations M is the molar mass and $K_{SV} = 1.343$ if V is measured in litres. $K_{SP} = 1.165$ if P is measured in atmospheres.

$$G = RT\ln\left(\frac{h^2}{2\pi\, mkT}\right)^{\frac{3}{2}}\frac{N}{V} = RT\ln\left(\frac{h^2}{2\pi\, mkT}\right)^{\frac{3}{2}}\frac{P}{kT}$$

$$= RT(-3/2\ln M - 3/2\ln T - \ln V + K_{GV})$$

$$= RT(-3/2\ln M - 5/2\ln T + \ln T + \ln P + K_{GP})\,.$$

Here

$$K_{GV} = 1.156 \text{ if } V \text{ is measured in litres}$$

$$K_{GP} = 3.657 \text{ if } P \text{ is measured in atmospheres.}$$

$$A = RT\ln\left(\frac{h^2}{2\pi\, mkT}\right)^{\frac{3}{2}}\frac{N}{V\mathrm{e}} = RT\ln\left(\frac{h^2}{2\pi\, mkT}\right)^{\frac{3}{2}}\frac{P}{KT\mathrm{e}}$$

$$= RT(-3/2\ln M - 3/2\ln T - \ln V + K_{AV})$$

$$= RT(-3/2\ln M - 5/2\ln T + \ln P + K_{AP})\,.$$

Here

$$K_{AV} = 0.156 \text{ if } V \text{ is measured in litres}$$

$$K_{AP} = 2.567 \text{ if } P \text{ is measured in atmospheres.}$$

APPENDIX 2: SUMMARY OF ROTATIONAL CONTRIBUTIONS TO THERMODYNAMIC FUNCTIONS

As in Chapter 4 we define, for a heteroatomic diatomic molecule

$$x = \frac{h^2}{8\pi^2 \, IkT}$$

$$q_{\text{rot}} = \sum_J (2J+1)\exp{-J(J+1)x}$$

$$= 1 + 3\,e^{-2x} + 5\,e^{-6x} + 7\,e^{-12x} + \ldots$$

$$\approx \frac{1}{x}\left(1 + \frac{x}{3} + \frac{x^2}{15} + \frac{4x^3}{315} + \ldots\right)$$

$$G_{\text{rot}} = A_{\text{rot}} = -RT\ln q_{\text{rot}}$$

$$\approx RT\left(\ln x - \frac{x}{3} - \frac{x^2}{90} - \frac{8x^3}{2835} + \ldots\right)$$

$$S_{\text{rot}} = R\ln q_{\text{rot}} + \frac{U_{\text{rot}}}{T}$$

$$= R\left(1 - \ln x - \frac{x^2}{90} - \frac{16x^3}{2835} + \ldots\right)$$

$$H_{\text{rot}} = U_{\text{rot}} = RT\frac{x}{q_{\text{rot}}}\left(6\mathrm{e}^{-2x} + 30\mathrm{e}^{-6x} + 84\mathrm{e}^{-12x} + \dots\right)$$

$$= RT\left(1 - \frac{x}{3} - \frac{x^2}{45} - \frac{8x^3}{945} + \dots\right)$$

$$C_{P_{(\text{rot})}} = C_{V_{(\text{rot})}} = R\left(\frac{x}{q_{\text{rot}}}\right)^2 12\mathrm{e}^{-2x}\left(1 + 15\,\mathrm{e}^{-4x} + 20\mathrm{e}^{-6x}\right.$$

$$\left. + 84\mathrm{e}^{-10x} + 175\mathrm{e}^{-12x} + 105\mathrm{e}^{-16x} + \dots\right)$$

$$= R\left(1 + \frac{x^2}{45} + \frac{16\,x^3}{945} + \dots\right).$$

APPENDIX 3: SUMMARY OF VIBRATIONAL CONTRIBUTIONS TO THERMODYNAMIC FUNCTIONS OF DIATOMIC MOLECULES

Let

$$u = h\nu/kT$$

$$q_{vib} = (1 - e^{-u})^{-1}$$

$$G_{vib} = A_{vib} = RT\ln(1 - e^{-u})$$

$$S_{vib} = R[u(e^u - 1)^{-1} - \ln(1 - e^{-u})]$$

$$H_{vib} = U_{vib} = RTu(e^u - 1)^{-1}$$

$$C_{P(vib)} = C_{V(vib)} = Ru^2 e^u (e^u - 1)^{-2}.$$

If u is small we may expand $e^u = 1 + u + 1/2u^2 + \ldots$ whence

$$G_{vib} = A_{vib} = RT\left(\ln u - \frac{u}{2} + \frac{u^2}{24} + \frac{u^4}{2880} + \ldots\right)$$

$$S_{vib} = R\left(1 - \ln u + \frac{u^2}{24} - \frac{u^4}{960} + \ldots\right)$$

$$H_{vib} = E_{vib} = RT\left(1 - \frac{u}{2} + \frac{u^2}{12} - \frac{u^4}{720} + \ldots\right)$$

$$C_{P(\text{vib})} = C_{V(\text{vib})} = R\left(1 - \frac{u^2}{12} + \frac{u^4}{240} - \dots\right).$$

Thus tables of vibrational contributions may be prepared, such as the following:

Thermodynamic functions of a simple harmonic oscillator

$u = h\nu/kT$	C/R	S/R
0.1	0.9992	3.3030
0.2	0.9967	2.6111
0.3	0.9925	2.2078
0.4	0.9868	1.9230
0.5	0.9794	1.7035
0.6	0.9705	1.5257
1.0	0.9207	1.0406
2.0	0.7241	0.4584
3.0	0.4963	0.2179
4.0	0.3041	0.0925
5.0	0.1707	0.0403
6.0	0.0898	0.0171

(taken from Mayer and Mayer, John Wiley, 1940).

APPENDIX 4: EQUILIBRIUM CONSTANTS

THE definition of Helmholtz Free Energy (Chapter 1) is

$$A = U - TS$$

but in Chapter 3 we showed that, in general:

$$S = k \ln Q + U/T.$$

Thus

$$A = -kT \ln Q.$$

For molar quantities of gases (which are non-localized particles), we saw in Chapter 4 that

$$Q_{\text{molar}} = q^N/N!$$

Therefore

$$A_{\text{molar}} = -kT \ln Q_{\text{molar}} = -kT \ln (q^N/N!),$$

but by definition (Chapter 1)

$$G = H - TS, \text{ and remembering that } H = U + PV,$$
$$G = A + PV.$$

Therefore

$$G_{\text{molar}} = A_{\text{molar}} + PV_{\text{molar}}.$$

For 1 mol of an ideal gas in its standard state ($P = 1$ atm.) the Gibbs Free Energy has its standard state value $G_{\text{molar}}^{\ominus}$ and $V_{\text{molar}} = RT/P = RT/1$.

Therefore

$$G_{\text{molar}}^{\ominus} = -kT \ln \left[(q^{\ominus})^{N}/N!\right] + RT.$$

Here $q^{\ominus}$ is the partition function appropriate to the standard state. By Stirling's approximation:

$$\ln N! = N \ln N - N$$

or

$$N! = (N/e)^{N}.$$

Therefore

$$G_{\text{molar}}^{\ominus} = -NkT \ln \left[(q^{\ominus}/N)e\right] + RT$$
$$= -RT \ln (q^{\ominus}/N) - RT \ln e + RT,$$

remembering that $\ln e = 1$

$$G_{\text{molar}}^{\ominus} = -RT \ln (q^{\ominus}/N).$$

We saw in Chapter 1 that

$$\Delta G^{\ominus} = -RT \ln K_{p}.$$

Thus the partition functions for standard states are related to equilibrium constants in terms of partial pressures, because $\Delta G^{\ominus} = (G_{\text{molar}}^{\ominus})_{\text{products}} - (G_{\text{molar}}^{\ominus})_{\text{reactants}}$.

If we consider a simple reaction of the type

$$A \rightleftharpoons B$$

$$\Delta G^{\ominus} = -RT \ln K_{p} = -RT \ln (q_{B}^{\ominus}/N)/(q_{A}^{\ominus}/N)$$

However, we must remember that in addition to the difference in partition functions there may also be a difference between the energies of the lowest levels of A and B. As we saw in Chapter 5 we take account of this difference $(E_{B}^{0} - E_{A}^{0} = \Delta E^{0})$ by including a term $\exp(-\Delta E^{0}/RT)$ in the expression for the equilibrium constant.

Thus in this simple case

$$K_{p} = \frac{(q_{B}^{\ominus}/N)}{(q_{A}^{\ominus}/N)} e^{-\Delta E^{0}/RT}.$$

For a general reaction

$$aA + bB \rightleftharpoons cC + dD$$

$$K_p = \frac{(q_C^\ominus/N)^c(q_D^\ominus/N)^d}{(q_A^\ominus/N)^a(q_B^\ominus/N)^b}e^{-\Delta E^0/RT} \, .$$

This result should be compared with the expression given in Chapter 5.

Example

For the equilibrium between sodium atoms and diatomic sodium molecules,

$$Na_2 \rightleftharpoons 2Na$$

$$
\begin{aligned}
K_p &= \frac{[q^\ominus(Na)]^2}{q^\ominus(Na_2)} \cdot \frac{1}{N} \exp\left(-\frac{\Delta E}{kT}\right) \\
&= \frac{(2\pi m_{Na}kT)^3/h^6}{(2\pi m_{Na_2}kT)^{\frac{3}{2}}/h^3} \cdot \frac{RT}{N} \cdot \frac{\sigma h^2[P_0^2(Na)/P_0(Na_2)]}{8\pi^2 IkT} \\
&\quad \times \left(1 - \exp\frac{-h\nu_0}{kT}\right) \exp\left(-\frac{\Delta E}{kT}\right) .
\end{aligned}
$$

Here

$$m_{Na} = \frac{1}{2}m_{Na_2} = 23/(6.02 \times 10^{23})\text{g}$$

I is moment of inertia for $Na_2 = \mu r^2 = (1.91 \times 10^{-26})(3.078 \times 10^{-10})^2$ m^{2}kg, ν_0 is vibration frequency and $\sigma = 2$.

$$\frac{h\nu_0}{kT} = \frac{6.62 \times 10^{-27} \times 3 \times 10^{10} \times 159.23}{1.38 \times 10^{-16} \times 10^3}$$

$P_0(Na)$ is statistical weight of ground state Na atom which is 2 since the spectroscopic state is $^2S_{\frac{1}{2}}$, owing to the unpaired 3s electron. $P_0(Na_2)$ is 1 as the molecule has no resultant spin or orbital angular momentum.

$$\Delta E = 0.73 \text{ eV}$$
$$= 0.73 \times 1.602 \times 10^{-12} \text{ J} \, .$$

When these numbers are substituted in the expression for K_p a value of $K_p \approx 2.5$ results.

PROBLEMS

Chapter 1

1.1. The vapour density of acetic acid at a constant pressure of one atmosphere varies with temperature as shown in the table. Use these results to calculate the enthalpy change of dimerization

Vapour density	T/K
58	405
42	477
32	556
31	593

1.2. At 1000 K and a pressure of one atmosphere the degree of dissociation of carbon dioxide to carbon monoxide and oxygen is 2×10^{-5} per cent. Calculate the standard free-energy change of the reaction, $\Delta G^{\ominus}$.

1.3. Calculate ΔU, ΔH, ΔA, ΔG, and ΔS when 1 mol of an ideal gas expands isothermally (at 300 K) from a volume of 100 dm^3 to 200 dm^3.

Chapter 2

2.1. The energy levels E_n of hydrogen-like atoms are given by the formula $E_n = -R/n^2$: here $R = \mu e^4/8\varepsilon_0^2 ch^3$, μ being the reduced mass

$m_e m_N/(m_e + m_N)$ with m_e the mass of the electron (9.1090×10^{-28} g) and m_N the mass of the nucleus. Given that the mass of a proton is 1.6725×10^{-24} g and that for H, $R = 109\ 678$ cm^{-1} calculate and plot the energy-level diagram for the isotopes hydrogen, deuterium, and tritium.

2.2. In how many ways can ten numbered balls be distributed into five boxes? If the balls are identical and not distinguishable by being numbered, how many distributions are then possible?

2.3. Find by tabulation the number of ways in which 4 distinguishable particles can be distributed among 3 energy levels. What is the effect on the number of distributions of (i) making the particles indistinguishable (ii) limiting the number of particles to two per level?

2.4. A sample of partially deuterated ammonia is analysed and is found to contain equal molar quantities of hydrogen and deuterium. Assuming that the distribution is purely random calculate the proportions of the molecules NH_3, NH_2D, NHD_2, and ND_3.

2.5. Test Stirling's approximation for a series of increasingly large numbers to assess at what sort of values of N the approximation is safe. Compare values of $N!$ found using Stirling's formula with those found using the more accurate formula

$$N! = (2\pi N)^{\frac{1}{2}} (N/e)^N \{(1 + 1/(12N) + 1/(228N^2))\}.$$

Chapter 3

3.1. Consider a system of N particles with only two energy levels separated by an energy gap ε. What is the maximum value of Q possible, and what is the minimum value? Under what conditions would these values be achieved? Derive values of Q for the cases where $\varepsilon = 0.1\,kT, kT$, and $10\,kT$.

3.2. Derive a general formula for the heat capacity C_V of a system which has two non-degenerate energy states separated by an energy ε (see Chapter 7 for an amplification of this situation). Are there any physically interesting situations where two-level systems are important?

3.3. Calculate the fractional distribution of a system of N particles in a three-level system with levels of energy 0, 0.5 kT, and 0.6 kT. Calculate also Q and S. What happens when the upper levels become degenerate?

3.4. From the expressions given in the text relating the thermodynamic functions U and S to the partition function, derive general formulae for A, G, C_V, and C_P.

Chapter 4

4.1. If students in a class are to be graded A, B, C, or D for their year's work, what is the simple, statistical probability of achieving grade A for every year of a four year course? What are the chances of averaging grade B or better over the four years?

4.2. Using the formula for the molar entropy of a monatomic gas, calculate the molar entropies at 298 K for He, Ne, Ar, Kr, Xe, and Rn.

$$R = 8.314 \text{ J K}^{-1} \text{ mol}^{-1} \quad h = 6.626 \times 10^{-34} \text{ J s}$$
$$e = 2.718 \quad\quad\quad\quad \pi = 3.142$$
$$V = 22.41 \times 10^{-3} \text{ m}^3 \quad k = 1.381 \times 10^{-22} \text{ J K}^{-1}$$
$$N = 6.022 \times 10^{23} \text{ mol}^{-1}.$$

Plot the logarithm of the entropy against $m^{3/2}$. Comment on the result.

4.3. By comparing the entropy found from calorimetric data with the value calculated using the statistical thermodynamic formulae for a diatomic gas (see Appendices) calculate the entropy of crystalline nitrogen in the limit of $T \to 0$, using the following experimental data:

transition temperature $T_{\text{tr}} = 35.61$ K; $\Delta H_{\text{tr}} = 228.7$ J mol^{-1}

melting point $\quad\quad\quad T_m = 63.14$ K; $\Delta H_m = 720.2$ J mol^{-1}

boiling point $\quad\quad\quad\quad T_b = 77.32$ K; $\Delta H_b = 5571.5$ J mol^{-1}.

$$\int_0^{T_{\text{tr}}} C \, d\ln T = 27.13 \text{ J K}^{-1} \text{ mol}^{-1}$$

$$\int_{T_{\text{tr}}}^{T_m} C \, d\ln T = 23.36 \text{ J K}^{-1} \text{ mol}^{-1}$$

$$\int_{T_m}^{T_b} C \, d\ln T = 11.41 \text{ J K}^{-1} \text{ mol}^{-1}$$

$$\text{Internuclear distance of } {}^{14}N_2 = 0.1095 \text{ nm}$$
$$\text{Vibrational frequency of } {}^{14}N_2 = 2360 \text{ cm}^{-1}.$$

4.4. The ground electronic states of the halogen atoms are doublets. The spacings between the lower level ($^2P_{3/2}$) and the upper level ($^2P_{1/2}$) are respectively: F = 404 cm^{-1}, Cl = 881 cm^{-1}, Br = 3685 cm^{-1}, I = 7603 cm^{-1}. Using the formulae for a two-level system calculate the electronic contributions to the entropies of the atoms at temperatures 0 K, 100 K, 1000 K and 10,000 K.

4.5. Using the Bohr formula for the hydrogen atom,

$$r = \varepsilon_0 n^2 h^2 / \pi m e^2$$

where $\varepsilon_0 h^2 / \pi m e^2 = 5.292 \times 10^{-11}$ m, estimate the value of the quantum number n for which the average 'size' of the hydrogen atom would be comparable in magnitude to that of an average room.

Chapter 5

5.1. Calculate the equilibrium constant K_P for $I_2 \rightleftharpoons 2I$ at 1000 K and at a pressure of one atmosphere. The internuclear separation is 0.2667 nm and the vibration frequency is 214.36 cm^{-1}. The iodine atoms are in a ^2P state which has a spin-orbit splitting $^2P_{1/2} \leftarrow {}^2P_{3/2}$ of 7603 cm^{-1}. (It may be helpful to consult Appendix 4.)

5.2. Calculate the temperature at which the equilibrium ratio of *ortho* to *para* hydrogen is unity. Use the following formula [Dennison, *Proc. Roy. Soc. A*, **115**, 483 (1927)]

$$\frac{[ortho \text{ } H_2]}{[para \text{ } H_2]} = \frac{3\{3\exp(-2\theta_r/T) + 7\exp(-12\theta_r/T) + \ldots\}}{1\{1 + 5\exp(-6\theta_r/T) + \ldots\}}$$

where θ_r is the 'characteristic temperature' of H_2 and is equal to 85.4 K.

5.3. As a rough guide to the effect of temperature on reactions it is sometimes said that it is typical for a rise of 10 K to double or treble the rate. What range of activation energies does this imply for (i) a reaction taking place at room temperature (ii) a reaction taking place near 800 K?

5.4. Make an order-of-magnitude estimate of the kinetic isotope effect which might be observed for reaction in which the rate determining step is (i) the breaking of a P–H bond or (ii) the breaking of a C–N bond. Assume that the stretching frequency of the P–H bond is about 2400 cm^{-1} and that of the C–N bond is about 1300 cm^{-1}.

Chapter 6

6.1. Calculate the vibrational heat capacity of $^{35}Cl_2$ at 200 K and 500 K. The vibration frequency of $^{35}Cl_2$ is 565 cm^{-1}. What will be the total heat capacity at constant volume at each temperature?

6.2. The heat capacity of 1 mol of iodine vapour at a constant pressure of 1 atm varies with temperature as follows. Just above the boiling point the heat capacity is almost independent of temperature ($C_P = 4.5\,R$). At about 1020 K the heat capacity starts to rise and reaches a high value ($C_P \sim 31\,R$) near 1520 K. At higher temperatures the heat capacity falls to about 5.0 R. How can these observations be interpreted?

6.3. Calculate the room-temperature heat capacity of monatomic solids with the following Einstein temperatures: 100 K, 400 K, 2000 K.

Chapter 7

7.1. Calculate enough values of the heat capacity of a two-level system to plot the variation of $C_{2-\text{level}}$ with temperature for a system in which the upper level has a degeneracy five times the degeneracy of the lower level.

7.2. Calculate the magnetic entropy of a magnetically dilute paramagnetic salt for which $J = \frac{1}{2}$. What is the maximum amount of heat that can be liberated if this salt is subjected to a magnetic field of 1 tesla (10,000 gauss) at 1 K?

7.3. Calculate the molar heat capacity of NO at 120 K and 180 K, given that NO has a double ground electronic state, $^2\Pi$, with a separation between the lower and upper components $^2\Pi_{1/2}$ and $^2\Pi_{3/2}$ of 121 cm^{-1}. Compare these values with the experimental results.

$$C_P(120\text{ K}) = 31.4 \text{ J K}^{-1}\text{ mol}^{-1} \qquad C_P(180\text{ K}) = 30.5 \text{ J K}^{-1}\text{ mol}^{-1}.$$

7.4. The energies of various conformations of a disubstituted ethane CH_2X–CH_2Y can be plotted as a function of the rotation of one substituted methyl group with respect to the other. Stable conformations will be indicated by minima in the curve of internal energy against the torsional angle. Explain why a knowledge of the relative energies of these minima is insufficient to permit a calculation of the relative proportions of the various conformers. (See Farnell *et al., J. Theoret. Biol.* (1974).)

7.5. Discuss the reasons why liquid hydrogen has a lower boiling point than liquid deuterium. What other differences in physical properties can be observed between the two isotopic species?

7.6. Suggest some of the ways in which life would be altered if room temperature superconductors became available.

Chapter 8

8.1. Assuming that the frequency scale on Fig. 8.1 is linear use the spectrum to obtain the following parameters of $H^{35}Cl$ (i) the internuclear separation (ii) the force constant of the bond and (iii) the zero-point energy.

8.2. Explain why the spectra of HCl at three different temperatures, shown in Fig. 8.3 differ in appearance. Estimate the temperatures of the gas in the two lower spectra. (Note that the rotational fine structure has been blurred so that the band appears just an envelope. This happens at high gas pressures. Why?)

8.3. If the intensity of a beam of light is reduced to one quarter of its original value when traversing a block of a substance, what will be the intensity when it has passed through another identical block, H?

8.4. What intensity alternation is to be expected in the rotational Raman spectrum of $^{35}Cl_2$? (The ^{35}Cl nucleus has a spin, $I = 3/2$.)

8.5. What analogies may be drawn between laser action and superconductivity?

BIBLIOGRAPHY

Chapter 1

The individual book which relates to much of the subject matter of this book and which we have enjoyed to the extent of being stimulated to authorship is *The third law of thermodynamics* by J. Wilks (1961) Clarendon Press, Oxford. This book contains much fascinating and readable material.

Basic chemical thermodynamics by E. B. Smith (1990), Oxford, Univ. Press provides an excellent introduction to classical thermodynamics. More advanced texts include *Chemical thermodynamics* by E. F. Caldin (1961) Clarendon Press, Oxford, and *The principles of chemical equilibrium* (3rd edn) by K. G. Denbigh (1971) Cambridge University Press.

Chapters 2, 3, and 4

Greater detail and rigour are provided by the treatments of material in these chapters by G. S. Rushbrooke (1964) *Introduction to statistical thermodynamics*, Clarendon Press, Oxford; R. H. Fowler and E. A. Guggenheim (1949) *Statistical thermodynomics*, Cambridge University Press; and J. E. Mayer and M. G. Mayer (1940) *Statistical mechanics*, Wiley, New York. This last book gives the various partition functions in particularly convenient forms for practical application of the results. T. L. Hill, *An introduction to statistical thermodynamics*, Dover, New York

(1986). D. Chandler, *Introduction to statistical mechanics*, Oxford Univ. Press (1987).

Chapter 5

Among the wide variety of good, comprehensive texts on equilibria and kinetics we have particularly profited from reading *The structure of physical chemistry* (1965) by C. N. Hinshelwood, Clarendon Press; *Chemical kinetics* (2nd edn) by K. J. Laidler (1965), McGraw-Hill; *Theories of chemical reaction rates* by K. J. Laidler (1969), McGraw-Hill, and *Kinetics and mechanisms* (2nd edn) by A. A. Frost and R. G. Pearson (1961), Wiley.

Chapter 6

The admirable book by J. C. Kittel, *Introduction to solid state physics* (4th edn) (1971), Wiley, should be consulted by those who wish to pursue in this subject to a more advanced level. Ionic solutions are fully dealt within the *Oxford Chemistry Series* titles *Ions in solutions* (2): *An introduction to electrochemistry* by J. Robbins (1972), Clarendon Press, and *Ions in solution* (3): *Inorganic properties* by G. Pass (1973), Clarendon Press.

Chapter 7

A fascinating account of production of low temperatures is given by O. N. Lounasmaa, Scientific American, **221**, No. 6, 26 (1969).

Chapter 8

Molecular spectroscopy is dealt with at an introductory level by C. N. Banwell, *Fundamentals of molecular spectroscopy*, McGraw-Hill (1966), in more detail by G. M. Barrow, *Molecular spectroscopy*, McGraw-Hill (1962), and with complete authority by G. Herzberg *Spectra of diatomic molecules* (1950) and *Infra-red and Raman spectra* (1945) Van Nostrand, New York.

A short account of Lasers can be found in the book by D. L. Andrew (1990) *Lasers in Chemistry*, Springer Verlag.

Chapter 9

An elementary account of this topic can be found in the short book by G. H. Grant and W. G. Richards, *Computational chemistry*, Oxford Univ.

Press (1995). A more comprehensive coverage is given by M. P. Allen and D. J. Tildesley, *Computer simulation of liquids*, Oxford Univ. Press (1987).

Chapter 10

The classic work on information theory is *Science and information theory* (2nd edn) by L. N. Brillouin (1962), Academic Press, New York.

A most readable cautionary essay on the application of these ideas to biological systems has been published by P. B. Medawar (1967) in an essay on Herbert Spencer included in *The art of the soluble*, Methuen, London.

INDEX